America Burning Revisited

National Workshop – Tyson's Corner, Virginia

November 30 - December 2, 1987

 FEMA

Foreword

On November 30 through December 2, 1987, the U.S. Fire Administration/Federal Emergency Management Agency conducted a three-day workshop on AMERICA BURNING REVISITED in Tyson's Corner, Virginia. Individuals from major professional businesses and governmental organizations with an interest in fire protection were invited to attend the workshop. The participants included several individuals who participated in the original work of the National Commission on Fire Prevention and Control, completed in 1973, as well as current members of Congress with oversight responsibility for the U.S. Fire Administration and the National Fire Academy.

The purpose of the workshop was to achieve a consensus, if possible, on the nature of the current U.S. fire problem, review and comment of the progress against fire since the original America Burning Report of 1973, and to develop recommendations for local, State and Federal efforts to reduce further the life and property loss due to fire.

The U.S. Fire Administration planned to use the results of this workshop to review its current programs and to set priorities for future direction and activities. Although the reader will not find major new or dramatic problems, you will discover that there are significant changes in the fire environment today, major changes in fire department responsibilities, increased concern about fire fighter health and safety, and a growing awareness that public fire education must be greatly improved and expanded, particularly for high-risk populations; the very young and very old.

We hope that the information in this report will generate renewed interest and commitment to further reducing our Nation's fire problems.

U.S. Fire Administration

Introduction

In 1973, the presidentially appointed National Commission on Fire Prevention and Control published *America Burning*, its landmark report on the nation's fire problem. The report presented 90 recommendations for a fire-safe America. For the past 15 years, *America Burning* has served as a road map, guiding the fire service and the federal fire programs toward the goal of improving fire safety in the United States.

The original *America Burning* report made 90 recommendations in 18 chapters in the following general subject areas:

- the nation's fire problem;
- the fire services;
- fire and the built environment;
- fire and the rural wildlands environment;
- fire prevention; and
- a program for the future.

While much of the report and its recommendations remained valid and relevant, it was time to take a second look at *America Burning* and re-examine the progress made toward the goals and objectives stated in the report. Perhaps more importantly, it was time to make new recommendations that would reflect the changes in our society and environment since 1973, but still move toward a more fire-safe America.

As a result, the conference on 'America Burning Revisited" was convened in the suburbs of Washington, D.C., from November 30 to December 2, 1987.

Purpose

"America Burning Revisited" had a threefold purpose. First, conference participants were to reach a consensus about the status of, and trends in, America's fire problem. Second, they were to revisit *America Burning* by reviewing and evaluating the progress toward the report's 90 recommendations. Finally, the conference participants were to recommend guidelines for local, state and federal efforts to reduce the life and property loss from fire.

The U.S. Fire Administration (USFA) planned to use the results of this conference as the basis for establishing its program priorities for future activities. This meant that the fire protection leaders participating in "America Burning Revisited" were to have the opportunity to map out the future course of fire safety in this country.

— Workshop Structure —

The participants invited to "America Burning Revisited" came from major professional, business and governmental organizations with an interest in fire protection. This included representatives of the fire service, the building materials industry, fire protection engineering, burn treatment centers, testing laboratories, labor, academia, building code groups, associations of local government officials, the housing industry, the elderly, private industry, and local, state and federal government, including the U.S. House of Representatives.

James F. Coyle

The conference participants were divided into seven work groups coinciding with selected chapters from the original *America Burning* report. A listing of the task group titles and topics follows:

- **Task Force 1. The Nation's Fire Problem**

Topics included causes, social changes, prevention issues, federal role, data collection and reporting, burn treatment and the role of the insurance industry.

- **Task Force 2. The Fire Services: Operations**

Topics included all field operational matters, for example, current missions and services of the fire service, alternative service delivery mechanisms, incident command, tactics, training, equipment, apparatus, and fire fighter health and safety.

- **Task Force 3. The Fire Services: Management and Administration**

Topics included all headquarters staff responsibilities, for example, budgeting, financing, master planning, personnel (Fair Labor Standards Act, etc.) and productivity.

- **Task Force 4. Fire and the Built Environment**

Topics included current hazards, codes and standards, transport (aircraft, motor vehicle, marine, etc.), commercial and industrial fires, causes and remedies, how people die in fire, hazards through design, and product design.

- **Task Force 6. Fire and the Rural Wildlands Environment**

Topics included volunteer issues, urban/wildland interface, and forest and grassland fire protection.

Task Force 6. Fire Prevention

Topics included fire safety education, home and workplace fire safety, built-in protection, public and private sector cooperation, and special population protection.

Task Force 7. Preparing for the 21st Century

Topics included research needs, federal involvement, and private and public sector involvement.

Each task force was assigned four identical tasks. They were to review the issues as stated in *America Burning;* identify new problems, concerns and issues; develop alternative solutions; and recommend actions for consideration by all participants.

Conduct

The conference began with a welcome from Julius W. Becton, Jr., the director of the Federal Emergency Management Agency.

Participants then were given several in-depth background presentations to help form or strengthen their frames of reference. "America Burning - The Past" was discussed by Lou Amabili, director of the Delaware State Fire School and a former member of the National Commission on Fire Prevention and Control. The administrator of the U.S. Fire Administration, Clyde A. Bragdon, Jr., talked about "America Burning - The Present." Finally, Dr. John Granito presented "America Burning -The Future." (Summaries of the Amabili, Bragdon and Granito presentations appear in the beginning of Section V of this report.)

Julius W. Becton

After the conference charge by U.S. Representative Doug Walgren (D-PA), chairman of the House Science, Research

Congressman Doug Walgren

America Burning Revisited   5

and Technology Subcommittee, participants divided into task forces for session one on problem identification. (Excerpts of Congressman Walgren's speech appear throughout Section V of this report.)

Day two of the conference (session two) concentrated on comparing the problems identified in session one with progress made over the past 15 years on the recommendations contained in *America Burning.* Along with several general and reporting sessions, the participants, in their task forces (session three, part one), also began to work on solutions to those problems.

Day three began with session three, part two, during which the task forces finalized their solution strategies and prepared their final reports. Each task force then presented its final report in a concluding general session.

Jim Estepp

In addition to the presentations mentioned above, the conference participants heard luncheon speeches from U.S. Representative Curt Weldon (R-PA), chairman of the Congressional Fire Services Caucus, and Chief M.H. "Jim" Estepp of the Prince George's County (Maryland) Fire Department. Remarks by U.S. Representative Sherwood P. Boehlert (R-NY) were presented by his aide, David Golston. Mr. Bill Honsell, Executive Director of the International City Management Association also spoke. Excerpts of the Weldon and Boehlert speeches appear throughout Section V of this report.

Before beginning the report of the conference, however, it is appropriate to set the stage with the information in Section II, "The Current and Projected Future Fire Protection Environment."

This report was produced by Wilkins Systems, Inc. The authors were Colin A. Campbell and Lee Feldstein. The editor was Colin A. Campbell. Section II, "Current and Projected Future Fire Protection Environment," and Appendix C, "Status Report on the 90 Recommendations from America Burning," were written by Harvey Ryland.

Current and Projected

Future Fire Protection Environment

─────── Introduction ───────

T his report contains a general discussion of the current and projected fire protection environment in the United States. The report includes information on:

- changes in fire experience, problems, issues and conditions since *America Burning* (the report of the National Commission on Fire Prevention and Control was published in 1973);

- current fire losses, major problem areas, programs and general conditions which directly, or indirectly, affect the protection of life and property from fire; and

- projected future (i.e., 1990 - 2020) conditions and situations at the local, state and national levels which might impact on fire protection.

── **General Trends 1973 - 1987-**

There have been significant changes in the field of fire protection since 1973. For example:

- Fires and fire losses have declined generally over the period 1973 - 1987. There are some exceptions to this trend that are discussed in Section 3.

- In general, fire departments have experienced major changes in such areas as emergency response activity,

available resources, fire fighter duties, requirements of the Fair Labor Standards Act (FLSA), volunteer recruitment and retention, and liability for personal and departmental actions.

- The United States Fire Administration (USFA), National Fire Academy (NFA) and Center for Fire Research (CFR) of the National Bureau of Standards (now the National Institute of Standards and Technology [NIST]) were established and have been operating for 14 years.

- The Consumer Product Safety Commission, U.S. Forest Service, General Services Administration, Department of Housing and Urban Development, Department of Health and Human Services, and other federal agencies have conducted programs to reduce fire losses and improve the effectiveness of fire protection.

- Building and fire codes have been strengthened, especially involving smoke detectors and automatic detection and suppression systems.

- There is a greater use of exotic materials, including those which produce toxic gases when burned.

- Fire fighter health and safety has been improved through physical fitness programs, new breathing apparatus (and its mandatory use), and clothing and protective equipment.

Fire Experience

In 1975, there were an estimated 8,100 fire deaths[1], with 6,200 in 1985, a reduction in annual deaths of 1,900, or 23%. This decline in fire deaths is illustrated in Figure 1. The decrease is even greater when figured on a per capita basis, as shown in Figure 2. On acumu-lative basis, this decrease represents an estimated saving of 6,900 lives over the period 1975-1985.

Fire fighter fatalities also have been decreasing. The annual number of fire fighter deaths (including line-of-duty deaths resulting from activities other than fighting fires, for example, training, station duty and responding to non-fire incidents) over the period 1979-1985 is illustrated in Figure 3. These reductions appear to result from a decrease in fire fighter deaths caused primarily by asphyxiation, probably as a result of increased use of breathing apparatus. It should be noted that heart attacks are responsible for approximately 50% of fire fighter deaths. Approximately 50%-60% of fire fighter fatalities are related directly to fire incidents.

Figure 1.

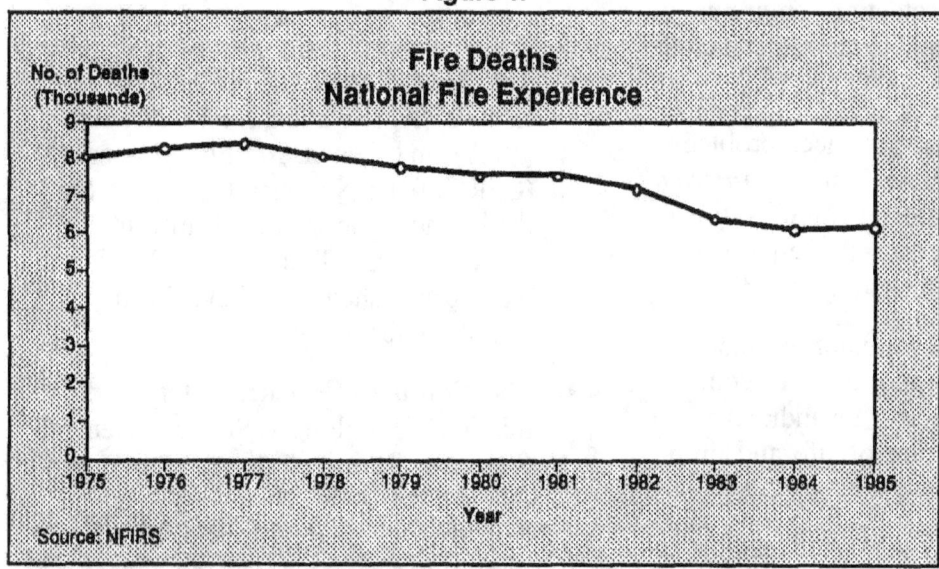

Figure 2.

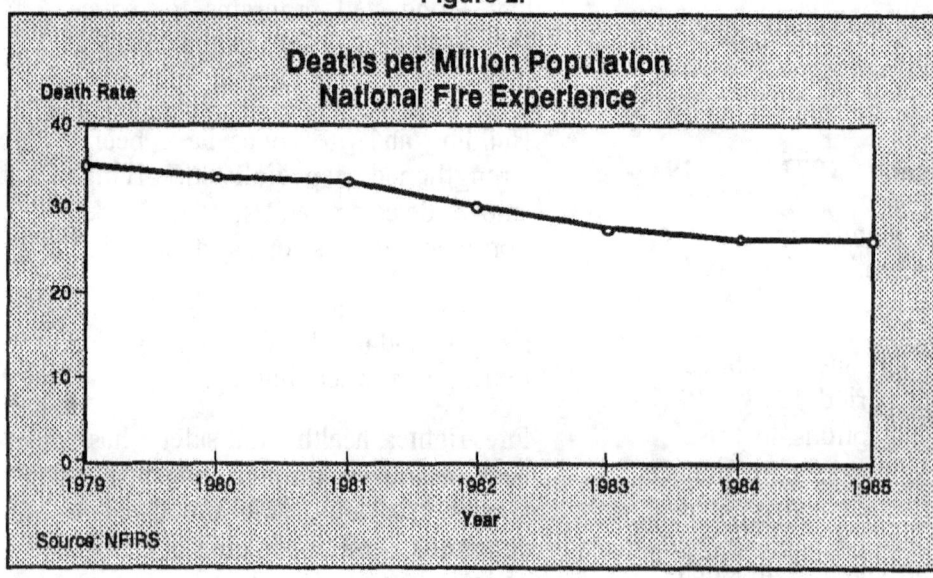

[1] All data used in this report were supplied by the National Fire Data Center, USFA.

The fatality rate for career fire fighters is 255 deaths per million workers, second only to mining which has a rate of 301 deaths. The rate for volunteer fire fighters is 25 deaths per million workers. However, the career fire fighters' risk of death cannot be compared directly to that of the volunteers on the basis of deaths per million workers. This is because most volunteer fire fighters do not work as fire fighters full time. Therefore, they are exposed to risk for less time than career fire fighters.

While fire deaths have been decreasing, fire-related injuries have remained relatively constant, even when considered on a per capita basis. The reasons for the lack of decrease in injuries are not known

Even greater reductions in fire deaths have been achieved within special categories. Clothing fire

deaths have fallen by 73% over the period 1968-1983, as shown in Figure 4. In addition, children's clothing fire deaths have dropped by 90%. These decreases are felt to be a result of flammable fabric standards developed and implemented by the National Institute of Standards and Technology and Consumer Product Safety Commission. Prior to the adoption of the

Figure 3.

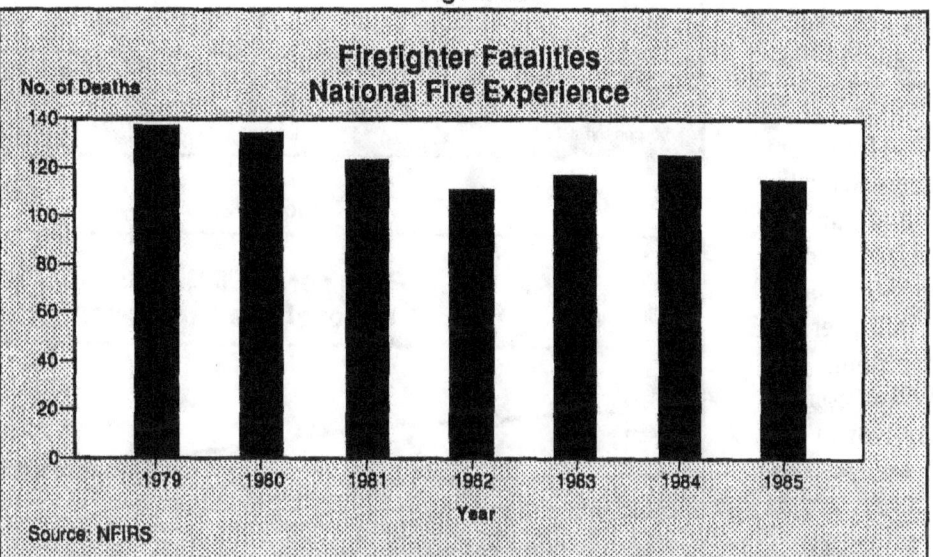

Figure 4.

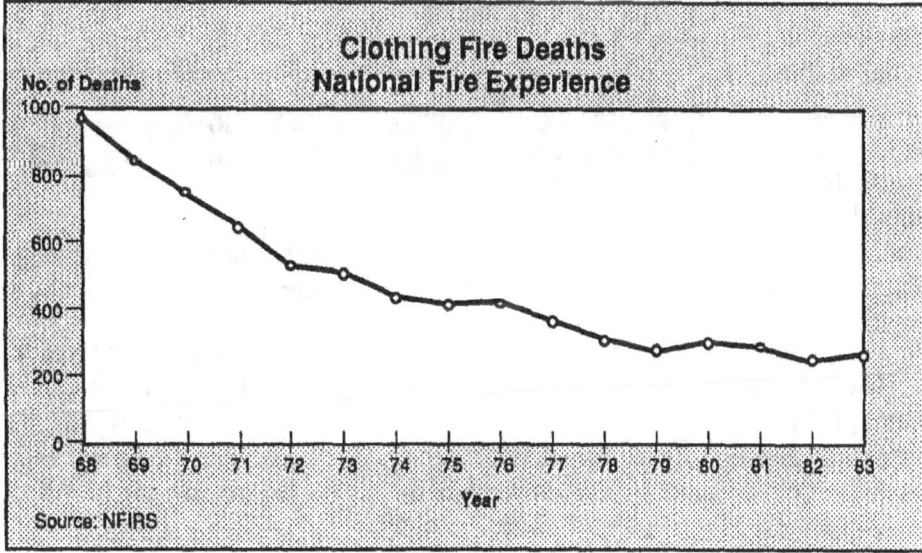

mandatory children's sleepwear standard, there were approximately 60 child sleepwear-related fire deaths per year; now there are approximately two such deaths per year.

However, even with these successes, clothing fires are still a serious problem. When considered on a risk basis, in 1985, there were 102 deaths per 1000 clothing fires, with 360 injuries per 1000 clothing fires. The next highest death rate is for portable local heater-related fires, with a rate of 37 deaths per 1000 portable local heater fires.

The mandatory mattress flammability standard has contributed to a 32% reduction in cigarette-ignited mattress fire deaths over the period 1980-1984. A voluntary standard for upholstered furniture has helped reduce cigarette-ignited upholstery fires by 24% over this same period.

Fires resulting from smoking in residential occupancies cause more fire deaths than any other single factor. This death rate is three to four times that of other causes. In 1985, 17.4% of all fire deaths were caused by smoking fires,

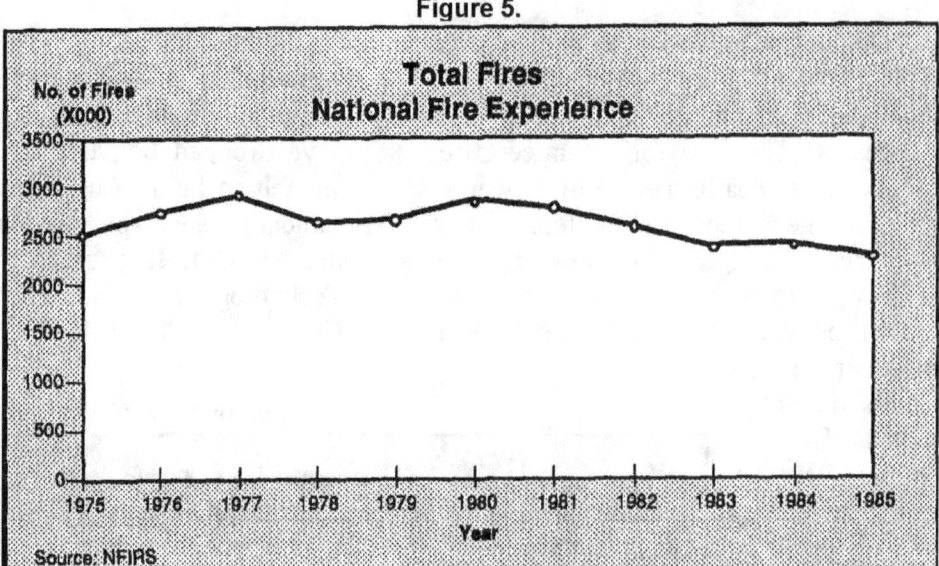

Figure 5.

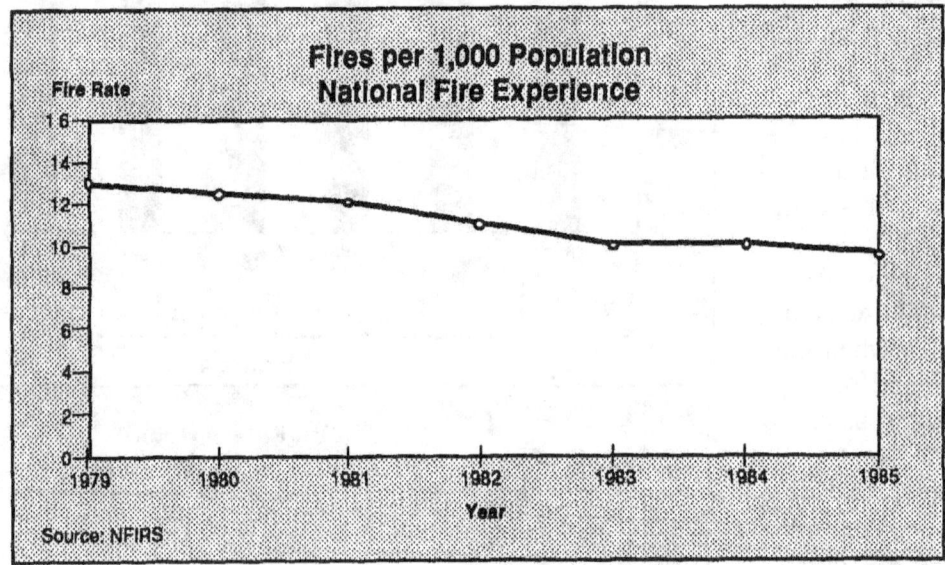

Figure 6.

Figure 7.

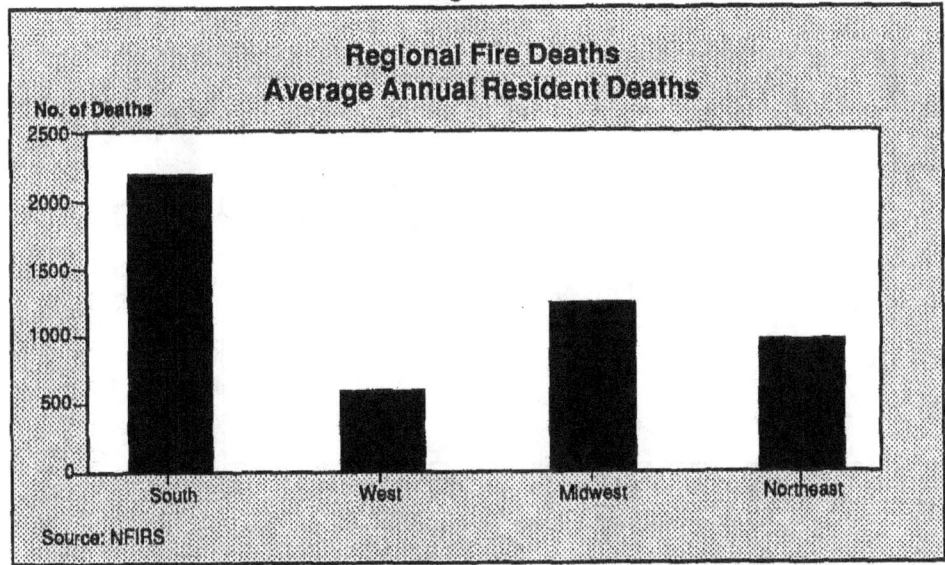

Figure 8.

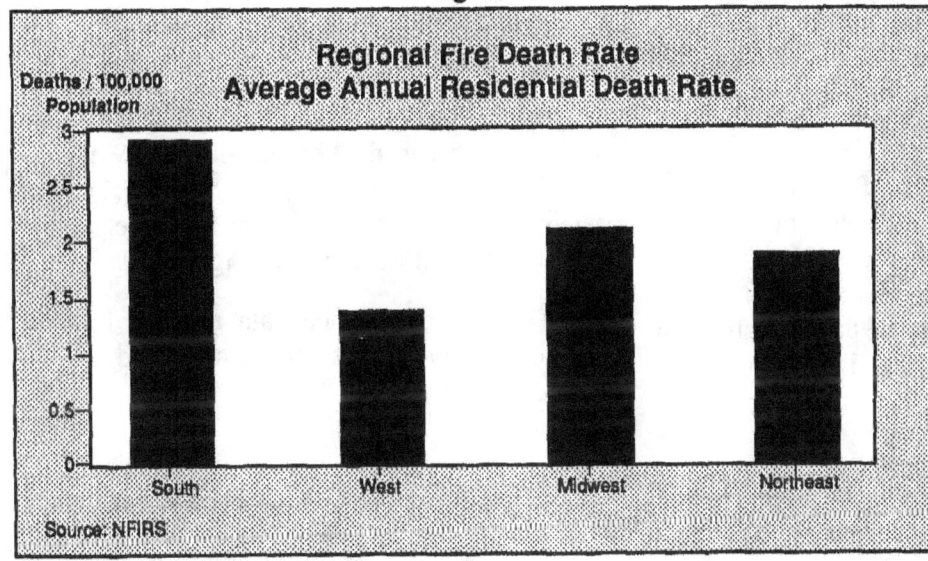

presented in Figure 5. The number of reported fires has been decreasing even though the population is increasing, which means that the number of fires per capita is declining, as illustrated in Figure 6.

Of course, the fire trend varies by area of the nation and type of fire. For example, the number of reported fires in California has declined *24%* over the 1975-1985 period. The residential fire death rate varies significantly across the nation. According to the Centers for Disease Control, the South has the highest annual average number of fire deaths (2150) and the highest death rate (29 per million population). By comparison, the West has the lowest number of residential deaths (585) and corresponding rate (14

with tire deaths caused by portable heaters second at 3.1%. Injury rates are also high for smoking fires - about twice that of other causes.[2]

The number of fires reported to fire departments has decreased by approximately 20% over the period 1975-1985, as

---

[2] This decline must be considered in light of the declining percentage of smokers in the country. According to the American Heart Association, the percentage of smokers declined from 43% to 32% over the period 1966-1983. *(Cigarette Smoking and Cardiovascular Disease,* American Heart Association, 1985)

Figure 9.

deaths per million population). The average annual number of residential fire deaths and per capita death rate for each region are illustrated in Figures 7 and 8.

Residential structure fires constitute only 25% of all fires, yet result in 74% of deaths, 62% of injuries and 43% of loss (1985 data). The fire-related death rates (deaths per 1000 fires - 1985 data) for the various types of residential occupancies are listed in Table I.

Fire deaths in rooming, boarding and lodging houses are clearly a major problem when considered on a per 1000 fire basis. However, in absolute terms, there are more fire deaths in one- and two-family dwellings and apartments.

The percentage of fires which are considered to be of incendiary or suspicious origin has been declining slightly, as shown in Figure 9. Because the total number of fires also has declined, this represents a decrease in the average annual number of arson fires. The percentages of arson fire-related deaths, injuries and dollar loss have remained fairly constant.

In actual dollars, the per capita fire loss has increased over the period 1979-1985. However, in constant dollars (1985 base), the loss rate has been relatively constant, as illustrated in Figure 10.

The fire experience data used in this section of the report are summarized in Table II.

— Fire Service Environment —

The fire service is going through a period of significant change. In fact, the degree of change is so great that it might be called "revolutionary" change instead of "evolutionary."

This change is felt to be primarily a result of three major factors: a) declining demand for fire suppression services; b) increasing demand for emergency medical services (EMS), hazardous materials protection, disaster preparedness and related

Figure 10.

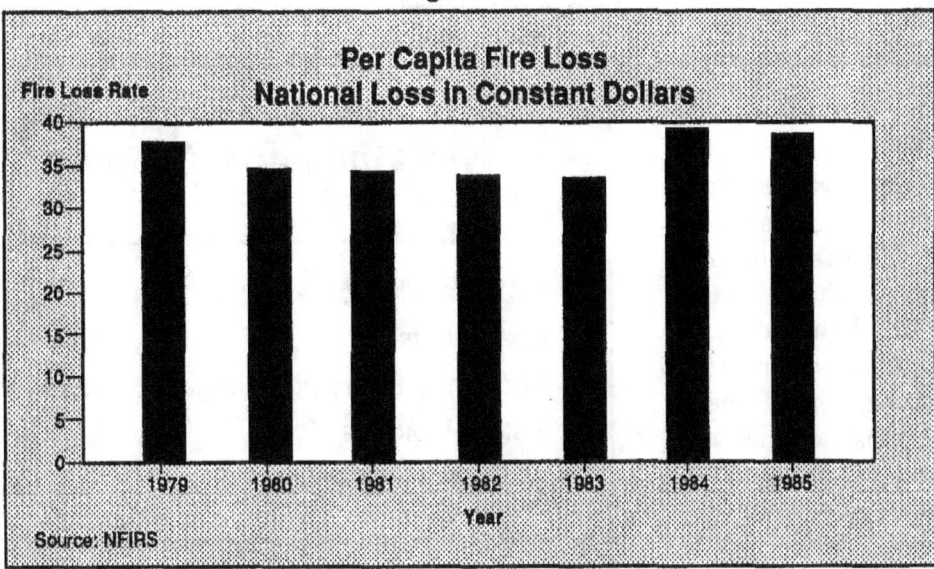

functions; and c) increasing competition for available local government resources.

Declining Demand for Fire
— Suppression Services —

As discussed in Section 3, the number of fires responded to by fire departments is declining generally throughout the nation. For example, fire incidents handled by the West Covina (California) Fire Department have decreased 24% during the period 1977-83, while the city grew ,by 40%.

Fires are not only going down in total, but also on a per capita basis (as shown in Figure 6). The exact reasons for this dramatic decrease in the number of fires have not been established quantitatively, However, they are felt to include a complex interaction of the following factors:

• improved fire prevention and public education programs conducted by fire departments and fire service personnel;

• improved building/fire codes;

• improved plan check and inspection programs conducted by fire prevention bureaus, as well as fire companies;

• increased use of smoke detectors and built-in suppression systems (which primarily reduce the extent of fire damage, injuries and deaths, but also can reduce the number of calls for service when detectors alert occupants to situations which then are handled without calling the fire department);

• improved public education programs for adults as well as juveniles;

• increased awareness of fire danger (for example, resulting from the MGM Grand Hotel fire);

• decreased numbers of people under age 25, thus reducing the number of individuals who might participate in juvenile firesetting; and

• improved national economy (however, if the economy declines, there may be a noticeable increase in the number of fires, particularly high-priced [e.g., gas guzzler] vehicles).

All of these factors (and others) probably will continue to reduce the number and magnitude of fires occurring in the future. In fact, the use of built-in detection and suppression systems is expected to continue to expand as new low-cost systems become available and as community administrators fully realize the power that such systems have to control the long-term costs of fire suppression. Not only will more and more construction include built-in suppression systems, but the trend to retrofit existing buildings is increasing. Several communities have passed retrofitting laws for specific occupancy types (hotels, for example), and several cities have considered retrofitting the entire community.

A number of jurisdictions have adopted fire sprinkler standards which require that all new construction (sometimes excluding single family dwellings) be fitted with a sprinkler system. Some communities require sprinklers in all new construction exceeding a given height or area.

Thus, it is considered probable that many communities will be protected by automatic detection and suppression systems at some time in the future. As a result, the demand for structural fire suppression will be limited to extinguishing fires in the room of origin, and incidents involving explosions, arson or other situations where automatic systems are turned off or defeated.

Table I.

Death Rates by Residential Occupance (Deaths per 1000 Fires)

1 & 2 Family Homes	8.1
Apartments	7.9
Rooming / Boarding / Lodging Houses	28.9
Hotels & Motels	9.1
All Other Residences	4.0

Even though many new and existing buildings will be protected by built-in systems, it probably will be a long time before structures in lower socio-economic areas (e.g., inner-city and rural neighborhoods) have such protection. Right now, those areas experience a disproportionate share of fires and related deaths. For example, the previously discussed high fire death rate for the South occurs in a region which has fewer smoke detectors in homes, higher usage of portable heating equipment and a larger percentage of persons below the poverty level. Thus, it will be very difficult to get such dwellings equipped with automatic detection and suppression systems.

Fire suppression services will be required for vehicle and wildland fires, as well as for such fire-related activities as evacuation/rescue, overhaul and salvage, and hazardous materials incidents. However, it is possible that the number of vehicle and wildland fires also will decrease in the future. Vehicles are being designed to be less susceptible to fire, and a number of programs are being consid-

Table II.

Summary of Fire Experience Data

Year	Fires (x000)	Fires / 1000 Population	Deaths	Deaths / Million Population	Firefighter Fatalities
1985	2,300	9.68	6,200	26.10	113
1984	2,400	10.20	6,200	26.30	123
1983	2,400	10.30	6,400	27.40	118
1982 [1]	2,613	11.30	7,200	30.20	110
1981	2,825	12.30	7,600	33.10	122
1980	2,890	12.80	7,600	33.60	133
1979	2,700	13.20	7,800	34.80	137
1978	2,690		8,100		
1977	2,960		8,500		
1976	2,780 [2]		8,400		
1975	2,600		8,100		

Year	Arson [7] (% of Fires)	Residential Deaths / 1000 Fires (Caused by Smoking)
1985	17.20	21.50
1984	15.80	30.00
1983	17.10	27.20
1982	18.30	27.60
1981	22.00	26.20
1980	21.00	24.70
1979	20.02	24.80

Year	Per Capita Loss (Constant $)	Injuries / Million Population	Clothing [3] Fire Deaths
1985	38.30	463	
1984	38.80	446	
1983	33.20	471	270
1982	33.50 [4]	425 [5]	260
1981	33.80	379	305
1980	34.20	393	311
1979	37.30	408	290
1978			325
1977			376
1976			426
1975			429
1974			445
1973			517
1972			542
1971			657
1970			750 [6]
1969			843
1968			977

1: Interpolated Data
2: Interpolated Data
3: Some Data May Be Incomplete
4: Interpolated Data
5: Interpolated Data
6: Interpolated Data
7: Includes Incendiary and Suspicious Fires

ered and/or implemented to minimize the consequences of wildland fires. For example:

- conducting prescribed burning to reduce fuel loading;

- converting fuel to less combustible types;

- creating greenbelts as fire breaks;

- restricting the use of combustible roofs; and

- requiring brush clearance around structures.

In addition, continued development reduces the amount of wildland acreage which is at risk. A contrary trend is that more structures are being built in the urban-wildland interface areas. Development in these areas means that more structures will be at risk and fire fighters increasingly will be faced with the dilemma of choosing between protecting exposed buildings or controlling the wildfire.

As a result of all of these factors, the overall demand for fire suppression services is expected to decrease in the future. However, the demand for fire *protection* will continue and possibly even increase as the effort is shifting to fire prevention and public education duties. Duties which are expected to increase (es-

- organizing and conducting public fire safety education classes; and

- creating and maintaining a high level of public awareness concerning fire prevention and fire safety.

Increasing Demand for — Emergency Medical Services —

The demand for emergency medical services (EMS) generally has been increasing in communities throughout the nation. It is not unusual for EMS incidents to constitute 80%, or more, of a fire protection agency's emergency responses. This demand is expected to continue to increase in the near term and then may level off or possibly even decline in the future. The increase is expected to be a result of heightened awareness of the availability of the service and the general aging of the population. The leveling off could occur because of generally improved health and "built-in protection." For example, pacemakers have been used for years, systems to internally inject medicine are in use and internal defibrillators soon will be available. Life-threatening calls for service may decline with these advances, but the greying of our population is likely to increase the per capita demand for emergency medical services.

pecially as the number and complexity of codes increases) include:

- reviewing proposed developments;

- reviewing construction plans (particularly for automatic detection and suppression systems);

- monitoring and inspecting construction and testing systems;

- conducting final inspections and issuing certificates of occupancy;

- conducting on-going inspections and corresponding codes;

Increasing Demand for Hazardous — Materials Protection —

The role of the fire service in hazardous materials protection has increased significantly over the past decade, and this role is expected to continue into the future. Hazardous materials incidents usually result from accidents that occur in the storage, transporting or processing of chemicals which endanger life or the environment, or lead to fire or explosion. Chemical spills associated with the transportation of hazardous materials (primarily truck or rail) have resulted in major disasters for even small communities and rural areas.

The severity of the hazardous materials problem from a national perspective is indicated by a statement in the *Hazardous Materials Planning Guide* (National Response Team publication NRT-1): "Recent evidence shows that the hazardous materials incidents are considered by many to be the most significant threat facing local jurisdictions."

The fire service originally became involved in hazardous materials because fire departments are usually first responders to most types of disasters. However, the fire department role has become more formalized as a result of local policies, and state and federal legislation.

At the federal level, the new legislation ("Emergency Planning and Community Right-to-Know Act of 1986" - Title III of the Superfund Amendments and Reauthorization Act) requires fire department involvement in a formal and extensive way. Under this legislation, fire departments are assigned a major role in preparing for, and responding to, hazardous materials incidents. Specifically, fire departments are designated to be represented on "Local Emergency Planning Committees" and to receive "Material Safety Data Sheets" (MSDS) containing information about hazardous materials in the community. The analysis, storage, retrieval and use of the MSDS is a responsibility for which many departments are not prepared, and may never have the resources to handle.

U.S. Representative Doug Walgren (D-PA)

"Even with the reduction in fire deaths, we still lose, if the loss could be dramatized, the equivalent of two fully loaded 747s colliding in mid-air every month - month after month after month."

Increasing Demand for Disaster ——— Preparedness ———

Fire departments are generally the first responders to almost every kind of disaster, including natural and technological hazards, and even domestic terrorism and foreign attack. In addition, many departments are using the Integrated Emergency Management System (IEMS) concept, which includes the fire service as a primary component of all four phases of emergency management: mitigation, preparedness, response and recovery.

Declining Community Resources

There are several factors which are interacting to limit community resources available to support the fire services. These factors are:

- taxing limitations (e.g., California's Proposition 13);

- spending limitations (e.g., California's Proposition 4);

- competition with other community services for resources which are available; and

- declining federal contributions (e.g., revenue sharing).

Competition for remaining funds is heavy, with social services constituting a major rival to the fire service. Moreover, it is predicted that the demand for social services will increase significantly in the future because of the demographic aspects of population, including the aging of the population.

Within the next 20 - 30 years, the average life span may be increased substantially. This would result in an older, larger population, causing a greater demand for community services. As a result, it is projected that communities will concentrate available resources on social services (e.g., housing, adult education, welfare, etc.) and will mitigate the demand for other services to the maximum extent possible. It is felt that fire protection offers the best opportunity for demand mitigation (among all community services) because of the success of automatic detection and suppression systems and fire prevention activities. It might be said that the fire service has been so

successful that it has reduced the demand for its services. In fact, there is nothing equivalent to a "sprinkler" for any other community service.

Community officials are beginning to realize the power that these systems have to reduce future costs. They will be taking actions to ensure that all new structures are equipped with sprinklers and then will move to retrofit existing buildings. In some cases, this will constitute a major policy shift. That is, in the past, administrators may not have supported automatic detection and suppression ordinances because of political pressure. However, it is felt that this position will change to enthusiastic adoption of such ordinances because it is an almost guaranteed way to achieve significant future cost avoidance.

The potential for long-range cost avoidance is not escaping the attention of community leaders, especially with decreasing per capita revenues and increasing demands for other services. Thus, there probably will be fewer additional fire fighters hired by fire departments than there would be under traditional forms of providing fire protection; and, these individuals probably will be performing a different mix of current duties, as well as assuming new responsibilities. Fire suppression services probably will always be required for rescue, evacuation, overhaul and salvage, as well as to fight those fires which are not extinguished by automatic suppression systems.

These changes will have a major impact on the fire services, and the anticipation and planning for this situation must begin now.

Regionalization

There is a general national trend toward the consolidation of fire protection services to save money, improve effectiveness and efficiency, and increase levels of service.

While quantitative information on the number of consolidations that have occurred could not be located, numerous examples of consolidation were identified in California, Colorado and Florida.

In addition to full jurisdictional consolidation, there are even more instances of partial consolidation, meaning the jurisdictions do not merge, but one or more services are shared or contracted out to another organization. Examples of partial, or functional, fire department consolidation are:

- communications/dispatch;

- apparatus purchase and/or maintenance;

- supplies purchasing and/or warehousing;

- fire cause and/or arson investigation;

- training;

- personnel recruitment;

- workers compensation and liability insurance coverage;

- fire prevention plan checking and inspection services;

- public education;

- data collection and analysis; and

- general management and administration.

Thus, the partial and full consolidation of fire jurisdictions is a component of the current fire protection trend which must be considered in planning for the future.

Summary of Trends

Fire protection is indeed in a period of significant change. Changes may occur which produce major shock waves, especially because of the traditional nature of the fire service. Yet, it is felt that these changes are probable and, in some cases, inevitable. At best, attempts to stop the changes only will delay what is occurring and might create situations where the fire service loses control of planning, policy development or even operations.

These changes do not mean the end of the fire service or even the end of fire suppression, They mean only that the type and level of effort dedicated to specific kinds of services will change, and to a greater extent in some communities. For example, the effort devoted to fire prevention (including public education) will increase, while the effort needed for suppression will decrease. Suppression services will be needed to respond to the fires that do occur, even in fully sprinklered buildings. However, in such structures, fire fighters seldom should have to fight a fire after it has reached flashover. The fires that do reach flashover probably will involve arson, explosion or other catastrophic failure. Thus, strong inspection/enforcement and arson control programs always will be needed (the determined arsonist likely will be able to defeat built-in systems). Fire suppression services also will be needed for wildland and vehicle fires.

The demand for emergency medical services is expected to increase over at least the foreseeable future. In addition, fire departments might assume additional duties which are extensions of more traditional activities. There is a trend developing for fire departments to assume responsibility for total building and environmental inspections. For example, the Environmental Protection Agency has determined that Radon gas can be a life-threatening problem in structures. If past trends continue, fire departments could be given a major role in dealing with this situation.

In aggregate, these changes will benefit the public because the protection of life and property will be increased significantly. To illustrate this conclusion, consider the case of the Disney World complex, a community with an area of approximately 43 square miles, day population of 200,000, night population of 25,000 and an assessed valuation of approximately two billion dollars. This complex is fully equipped with automatic detection and suppression systems. Over the last 15 years, the fire loss for Disney World has averaged $5,000 per year (including water damage), with no fire-related injuries or deaths.

Fire fighters also will benefit from these changes by having a safer working environment; job-related injuries and deaths should decrease significantly in the future because it is safer to fight a fire before it has reached flashover. Also, currently employed fire fighters will benefit by retaining existing jobs and compensation. It is almost certainly true that a growth in the number of fire fighter positions generally will not occur. Thus, it is in the best interests of the fire protection community to work together to recognize the changing environment and evolve to meet future requirements and conditions in the most effective and efficient manner. Moreover, no matter how good the projections of the future are, there always will be new challenges facing the fire protection community, situations that we cannot even conceive of at this time. For example, who could have predicted the occurrence of AIDS and the corresponding major impact on almost every element of society, including the fire service?

The tradition of the American fire service has been to accept these challenges and work on a dedicated, unselfish basis for their solution. This tradition will be continued as fire fighters meet future challenges using new and old techniques and resources.

Portions of this report were adapted from the document, "Feasibility Study of Fire Protection Cooperation in Santa Barbara County," prepared for the Santa Barbara County Firefighters - Local 2046, August 1987.

Executive Summary

Introduction

The conference participants were divided into seven task forces that were to discuss and analyze selected chapters from the original *America Burning* report. Those task forces were:

- **Task Force 1. The Nation's Fire Problem**

- **Task Force 2. The Fire Services: Operations**

- **Task Force 3. The Fire Services: Management & Administration**

- **Task Force 4. Fire and the Built Environment**

- **Task Force 5. Fire and the Rural Wildlands Environment**

- **Task Force 6. Fire Prevention**

- **Task Force 7. Preparing for the 21st Century**

In addition to a number of background and scene-setting presentations (summaries and excerpts of these speeches appear throughout Section V of this report), the task forces participated in four individual sessions during the three days of the conference.

It was the responsibility of the seven task forces to look at the original report's recommendations, study the program activity in the intervening years, and decide what really has been accomplished and what remains to be done. In other words, the participants were to answer the question, "Where do we go from here?"

The summaries of the seven task force reports that follow in this section give brief answers to that question.

Task Force 1- The Nation's Fire Problem

The theme that ran throughout the discussion among Task Force 1 members was "...Let's build on our successes to meet tomorrow's challenges." All agreed that the recommendations of *America Burning* are still valid and will continue to be in the coming years. With *America Burning* as a master plan, many successful programs have been delivered through federal, state and local government systems. Based on the successful track record ,of having a federal focal point to offer technical and financial assistance "when available," the task force felt the effort to control and reduce the fire problem in the United States would continue to be successful.

The problem of "fire" in the United States today can be summarized as follows:

- the US. has one of the highest fire death rates per capita in the industrialized world;

- each year, fires kill more Americans than all other natural emergencies combined, including floods, hurricanes, tornadoes and earthquakes;

- fire is the third leading cause of accidental death in the home; and

- the total annual cost to the American public from fires approaches $30 billion;

The characteristics of the nation's fire problem have changed little since the publication of *America Burning*. Eighty percent of our fire deaths continue to occur in residences. People living in rural areas and the inner city are two to three times more likely to die in fires in their homes than people in mid-size cities and suburban areas. The elderly and very young children are at the greatest risk in society. Fire disproportionately affects those people without a voice. It has much less effect on "the power structure." And the impact on a poor family that has a fire is more devastating.

As shocking as these statistics are, they do represent an improvement. The number of people killed by fires annually has decreased. However, during the very years we have pushed down the number of deaths per year caused by fires, the public, local officials and the fire service itself have yet to come to grips with the fact that America's "fire problem" may be a misnomer because it now includes emergency medical services, hazardous materials response and all those other

U.S. Representative Sherwood P. Boehlert (R-NY)

"This conference...must galvanize the entire fire service community to work together with Congress to preserve and expand the services of the federal fire programs. The entire Congress, all 535 U.S. Representatives and Senators, must understand the threat fire poses to their communities and the sacrifices made by the fire fighters of their communities."

"emergency" problems dealt with on a day-to-day basis by our nation's fire departments.

The status of America's fire problem, in this broader sense, is simply not known - not its magnitude, its major characteristics nor its victims. Data is not available.

Task Force 1 examined the nation's fire problem in this broader sense of the term and attempted to articulate the underlying causes and parameters of the total problem. In summary:

We have failed to convince elected officials of the seriousness of the fire problem. Fire service organizations need to find their common ground and present that with one voice. We need to support and educate the new Congressional Fire Services Caucus. Locally, fire departments need to establish citizens' committees or involve community groups in their work to gain support. They need to broaden the base of organizations having input to determine "the common ground." The public and officials may better appreciate the fire problem if it is explained to them in more brutal and vivid terms.

We need to understand and target such underlying causes as poor housing conditions and public apathy.

We need to fill major information gaps in the nation's fire problem, e.g., profiles of burn victims, EMS requirements, and hazardous materials requirements. The research areas identified are clear; the only requirement is money. The USFA and CFR have the legislative authority to do these studies, but lack money and personnel.

We need to redefine the fire problem. Fire service responsibilities have grown to include such other areas as hazardous materials and emergency medical response. Where proliferation of responsi-

bilities is coming from federal regulations, Congress needs to be more sensitive to the impact of these responsibilities on state and local governments.

The positive impact of smoke detectors needs to be maintained and extended to high-fire-risk population groups. Over the past decade, the fire service and the USFA have initiated successful programs to encourage the use of smoke detectors in new and existing residential occupancies. Public education programs for maintaining smoke detectors also have been proven successful. These efforts need to be continued.

Task Force 2 - The Fire Services: Operations

"How can fire protection be improved?" Both the National Commission on Fire Prevention and Control and the members of Task Force 2 replied that the answer lies not, in "more funds and more staff," but through improvements in the cost-effectiveness of existing service delivery. The objective of Task Force 2 was to review the progress toward improving the operational capability of the fire service since *America Burning* and to recommend new actions required to meet current and future demands.

The fire service is in a transitional period that calls for attention to varied and, at times, conflicting demands. The fire service needs to respond to the traditional, although diminishing, functions of fire suppression, while becoming more dynamic to meet new challenges. Code enforcement, emergency medical services and hazardous materials response are growing responsibilities that are being placed on the shoulders of the fire service.

The task force concluded that the primary fire protection issue from a fire service operational perspective is "the need to develop fire service human resources." Consistent management expectations and performance criteria for fire service personnel are generally lacking throughout the profession. Neither career nor volunteer services have fully addressed this problem, which is particularly acute at the officer level.

Part of the problem stems from a lack of nationally accepted minimum qualifications for all positions. Further, the recruitment and promotion of personnel is often idiosyncratic because of the lack of nationally accepted and validated selection procedures.

Personnel issues remain a significant problem for volunteer departments. There is a lack of credible information on volunteer membership trends, for example, why do volunteers join, remain active or quit?

The task force recommended a comprehensive research effort to analyze these issues. The study must be conducted professionally in order to be credible not only to the fire service, but to a broad spectrum of elected and appointed officials and peers in allied fields. Credibility is essential because it will address the dual problems of both effective and efficient fire protection service delivery, as well as the key areas of equal employment opportunity, funding and legislative activity.

Unless the fire service leadership enjoys more professional credibility, critical decisions affecting the fire service will be made more frequently by external actors (e.g., courts, elected officials) without the benefit of the fire service perspective. This problem affects both the volunteer and the career services equally.

Task Force 3 - The Fire Services: Management and Administration

Task Force 3's area of concern lay in the issues and recommendations discussed in Chapters 3 through 7 of *America Burning*. The objectives of the task force were to review the key managerial and administrative issues facing the fire service today and into the next century,

The task force participants emphasized that the problems facing today's fire service must be faced by both career and volunteer fire managers. The problems must be recognized by all fire departments regardless of size, location or type (career, volunteer or combination).

A theme that ran throughout the discussion was that the fire service frequently has the authority and resources to solve many of its own problems, and that many of these problems are primarily "people problems."

Some of the issues can be handled internally by the fire service. Continued support and financial encouragement is needed (a true partnership) from the U.S. Fire Administration to facilitate these changes.

The task force members felt that the "America Burning Revisited" conference must reinforce the positive accomplishments of the original *America Burning*, otherwise the report of the conference will not be complete. The many accomplishments of the past 15 years should be acknowledged and highlighted. However, many of the same problems discussed by the National Commission on Fire Prevention and Control in 1973 exist today, and some of these have grown worse over the years. In addition, the past 15 years have presented the fire service with new problems that were not anticipated in the original report. Therefore, the list of problems and action strategies should be even longer than in the original report.

The task force identified five major problem areas and developed solution strategies for four of them.

The top issue was fire and emergency service health and safety. Among the actions recommended to address this issue was the development of an insur-

ance/loss reimbursement program for fire fighter injury or loss of life to reimburse the jurisdiction. Fire departments need help when purchasing equipment or services. A testing or evaluation program similar to the Underwriters Laboratory certification system should be developed. The duplication of "right-to-know" legislation among state and federal governments should be avoided. AIDS and other communicable diseases are a threat to emergency service personnel and information should be made available about this problem.

The second major problem area was the lack of adequate standards to evaluate the delivery of fire department services.

Effectiveness and efficiency measures for fire and emergency service agencies should be developed. Further, the Integrated Emergency Management System (IEMS) concept should be the centerpiece of fire service planning and operations.

Personnel management was another problem area identified by the task force. Productivity and innovation need to be emphasized in all personnel management practices. Continuing education for all fire officers is a priority, as is the aggressive implementation of equal employment opportunity programs in both career and volunteer fire departments.

The fire service should conduct more concise and comprehensive planning, according to the task force. The USFA should conduct a study on alternative revenue sources, with special emphasis on user fees to better distribute the cost of the fire service between the public and private sectors. Research should be encouraged on improved fire-safe designs and building materials. Master planning

must be reinstated as a major program of the USFA and NFA. Finally, any federally or state-required responsibility must be accompanied by adequate funding.

Task Force 4 -
Fire and the Built Environment

The objectives of Task Force 4 were to define the fire protection problem vis-a-vis the built environment, to assess the progress that has been made since the publication of *America Burning* and to recommend strategies for improving building fire safety. The task force based its deliberations on Chapters 9 through 12 of *America Burning*. The scope of its discussion was broad and included the fire hazards posed by building design, the materials used to construct and furnish buildings, occupant behavior and use, building code development and enforcement, and the production, use, storage and transportation of hazardous materials.

The task force's discussion on fire safety in the built environment focused on problems related to public behavior, technology and code enforcement. *The public's lack of awareness of the fire problem is a key obstacle to reducing the nation's fire loss.* The public needs to be made part of the solution. Currently, there is little realization on the part of the public regarding the personnel cost and business consequences of fire. The task force outlined a national fire prevention education plan to address this problem. The strategy makes use of all media and is based on surveys and objective case studies of existing programs. It is essential that the message be vivid and brutal. Young children and teenagers need particular attention.

Incentives to encourage safe behavior and modify, hazardous behavior need to be instituted. Tax incentives, insurance credits, and legal and social deterrents were suggestions for modifying behavior.

Gaining control over the building contents is another key avenue for enhancing building safety. One approach involves greater control over a building's occupancy and use. More vigilant code enforcement is required to accomplish this task.

In addition, the fire service needs more information on measuring the combustion and toxicity characteristics of building contents. Test methodologies need to be developed to evaluate a product's ignitability, fire growth and toxicity. This information then must be made available to all end users of the product. We also need data on suppression systems and how a room performs during fire. The Center for Fire Research fire hazard analysis for single-family homes needs to be expanded. Once this information is available, it can be used to prevent flashover and design sprinkler and detection systems and fire barriers.

The task force lamented that specialized facilities and educational programs to train fire protection engineers are decreasing. While more graduate and doctoral programs are needed, the profession first needs to develop uniform curriculum and scholarship criteria.

Task Force 5 - Fire and the Rural Wildlands Environment

Task Force 5 concentrated on the issues and recommendations discussed in Chapters 13 and 14 of *America Burning*. The group's task was to evaluate the issues facing rural and wildland fire protection in light of current and future requirements for controlling fire in these environments.

The highest priority problem is public education, or more accurately, the public's lack of awareness of the fire problem. The fire service needs to market itself and its mission more effectively. It must use the latest techniques and the media to develop programs targeted at such key groups as the general public, elected and/ or appointed officials, and peers in related professions.

Local governments need to plan and control land use to take into account the effect of development on water supply, access roads and other fire hazards in the nation's wildlands. As more of the wildland areas become urbanized, the greater is the need for fire prevention and control master planning.

The fire service has too little input into the development of building and fire codes. Moreover, codes often are designed for cities and are not always appropriate

for rural and wildland areas that are characterized by scarce or inadequate water supplies, longer response times and volunteer fire departments. Standards and codes specifically addressing the problems of the rural and wildland environment need to be adopted. A national sprinkler standard appropriate for rural use would be valuable.

Rural departments frequently have problems taking advantage of state, regional and national training opportunities. The lack of funds and the unique scheduling constraints of volunteer departments impede access to necessary training. Legislation should be introduced to strengthen state and national training programs and to secure the stipend program at the National Fire Academy. The scheduling of training programs must be arranged so volunteers can attend. A training needs assessment system should be developed to ensure that programs concentrate on the issues and needs of the rural/volunteer fire service. Training is needed in arson control, the incident command system and fuel build-up forecasting.

Fire protection in rural and wildland areas often involves interagency cooperation. *Unfortunately, even with the models provided through FIRESCOPE and the Integrated Emergency Management System (IEMS), communication among agencies both within and across jurisdictions remains poor.* New work needs to be done in the area of inter- and intra-jurisdictional relationships. Action plans to control major fires and conflagrations would benefit from this improved coordination among key actors.

Fire service organizations need resources to deliver required services. These necessary resources include funding, community commitment, equipment and personnel. The task force felt that all of these elements needed additional support. If fire prevention and control services are going to meet the requirements set by the community, the fire service needs the tools to accomplish the job. The task force identified some options for acquiring the necessary tools, for example, a local and state group purchasing plan, and government grants and low-cost loans.

Task Force 6 - Fire Prevention

The National Commission on Fire Prevention and Control noted that fire prevention was the key to effective and efficient fire protection services. Fire prevention activities will become even more important in the future if fire department resources continue to decline because "..it is cheaper to prevent a fire than to fight it."

Much has been accomplished since 1973. Many of the nation's fire depart-

ments have significantly increased fire prevention activities over the past decade. In-service company fire safety inspections and participation in public education programs now are common. Fire departments have enlarged their role in supervising the fire safety of the built environment. They are more involved in overall land use planning and reviewing plans for new construction. Fire service officials frequently are involved in the development and adoption of model fire codes and the acceptance of local amendments to strengthen these codes, for example, codes requiring installation of smoke detectors in existing dwellings.

Upon reviewing the progress made over the past decade, the task force identified two primary problem areas that need to be addressed. *First, the public's awareness of the fire problem remains a major obstacle.* As a result, fire prevention still is not perceived as a high priority. Industry, political leaders and the fire service itself need to have a greater appreciation of fire prevention.

To overcome this obstacle, a national education campaign should be instituted.

U.S. Representative Doug Walgren (D-PA)

"I hope that you would ask yourselves whether we need a new organization that has a structure within which could fit all the elements of our society that are active in fire prevention...There has to be some effective continuous process that would include fire service organizations...citizen advocacy groups.the various non-profit organizations that have developed and served so well in the American system...business interests that have so much to lose...educators that have such a role to play in education...and technologists who could add, particularly in education, what new technology might provide...I hope you will think about how you can give congress the kind of direction we obviously need given the nature of the Congressional process, and I hope you will think about how you can give each other the ongoing organizational structure that would enable us to pull everybody into a common effort that may not be quite what they are interested in or may be contrary to some of their immediate interests on a political level..., but nonetheless can result in a greater unity of effort in the next 10 years than we have seen in the last 10."

The task force recommended that the campaign should build on such existing, successful programs as the National Fire Protection Association's "Learn Not To Burn" project. The fire service must join with industry leaders (e.g., the National Advertising Council) to launch a nation-wide public relations effort. Fire service support of the Congressional Fire Services Caucus is an effective way to educate political leaders about the fire problem. Fire prevention needs to be institutionalized in the fire service. Fire prevention training and active experience should be required of all chief officers.

> *"U.S. Representative Curt Weldon (R-PA)*
>
> *. ..What we can do is begin a process - begin a process of bringing in members who are willing to make a commitment, a real commitment, a genuine commitment, to the problems and needs of the fire service and the fire problem in America."*

The second key problem area is the lack of information needed to analyze the fire problem and evaluate programs. The task force commented that the current national data system(s) is inadequate. The information comes from a variety of sources, and quality control over data collection/entry is questionable. The task force acknowledged the wealth of data being compiled through the National Fire Incident Reporting System (NFIRS), but, took the U.S. Fire Administration (USFA) to task over the NFIRS lack of utility. "The information is too little and too late." The fire service desperately needs more analysis of the fire problem.

The only solution to the data problem is for the USFA to reinstitute a data

collection and analysis effort similar to, if not greater than, the one initiated with the launching of NFIRS. The data program needs to coordinate its data collection efforts and develop products that are accessible and useful to the fire service at both the national and local levels.

Task Force 7 -
"Preparing for the 21st Century"

The thrust of Task Force 7's mission was to recommend alternate concepts, policies and programs that would assist the fire service in planning and preparing for the future. The task force based its review on the issues and recommendation presented in Chapters 18-20 of *America Burning*.

The task force identified five key issues that need to be addressed as the fire service prepares for the 21st century. The issues are ranked-in priority.

Recognizing that fire safety's roots lie in the socio-cultural fabric of society, the task force considered the public's cultural orientation to fire as the primary critical issue for the next century. Attitudes, behavior and values contribute to this nation's relatively poor fire safety record as compared with other industrialized countries. Fire is much more salient in other cultures, where there may be rewards, as well as penalties, associated with fire safety behavior.

A campaign needs to be launched to change the public's cultural orientation toward fire. The campaign needs to be sophisticated and use state-of-the-art techniques in behavior modification and marketing.

Fire safety is a political issue. Laws, regulations and funding result from political efforts. The task force members felt that today's fire service was fragmented, and it could represent itself to Congress and other interests in a more effective manner. A nationally organized and coordinated capability to identify overall problems, establish priorities and "make things happen" does not exist. Further, the fire service primarily talks to itself. It needs to organize with other allied interests.

A fire safety political action plan needs to be organized. It should consist of the following three elements: first, support for the Congressional Fire Services Caucus; second, a "conference board" consisting of representatives from all fire service groups and other allied interests to lobby on behalf of fire safety issues; and, third, a political action committee.

We are still only beginning to understand the technics office. The physics and chemistry of combustion, toxicity and suppression are, in many respects, still an enigma. While we have gained a lot of ground over the past 15 years, much more needs to be learned about the phenomenology of fire and about the hazards and risks posed by building materials and furnishings. Research needs to be initiated to develop a new generation of detection and suppression technology.

The fire protection community needs to have information about the fire problem and how to manage it. The data collection and analysis conducted by the fire community are fragmented and incomplete. The U.S. Fire Administration has made progress with NFIRS, but not all states and cities are participating. Fur-

ther, other types of information are needed, as are better dissemination methods. The federal resources committed to this effort need to be augmented.

A national fire information database network should be established that includes not only fire incident statistics, but a sharing of ideas and experiences, as well.

The role and responsibilities of the fire service are being redefined as the environment and the demands of local government change. Traditional fire department activities are being replaced or supplemented with such new challenges as emergency medical services and hazardous material response. The budget crises faced by governments over the past decade have affected the resources available to the fire service, causing fire departments to reevaluate the services they provide and how these services are provided. Research needs to be conducted on new fire department organizations, responsibilities and working relationships.

Summary

The discussion and recommendations from this conference underscore the fire service's increasing professionalism. The participants did not hesitate to examine critically how the fire service goes about its job. While it is important to evaluate the constraints and requirements that are imposed by an organization's environment, the task forces did not shy away from issues that many would consider very close to home. Whether the topic involved fire service personnel selection and development standards, management priorities, new and untraditional activities, or new technologies, the participants evaluated how well the fire service was prepared to meet current demands, as well as those anticipated in the future.

Common themes included the need for more and better information on the dynamics of fire, life safety and emergency management issues. The fire service needs to be more active in controlling its destiny. This requires organizing at the local, state and federal levels to pursue its interests and objectives. The fire service needs to take an active stance in selling fire safety to the public, government officials and industry. Finally, the fire service needs to ensure that it has the proper personnel, resources and organization to deliver public safety services and manage inevitable future change.

Workshop Task Force Reports

Introduction

The conference participants were divided into seven task forces that were to discuss and analyze selected chapters from the original *America Burning* report. Those task forces were:

- **Task Force 1. The Nation's Fire Problem**

- **Task Force 2. The Fire Services: Operations**

- **Task Force 3. The Fire Services: Management and Administration**

- **Task Force 4. Fire and the Built Environment**

- **Task Force 6. Fire and the Rural Wildlands Environment**

- **Task Force 6. Fire Prevention**

- **Task Force 7. Preparing for the 21st Century**

In addition to a number of background and scene-setting presentations (summaries and excerpts of these speeches appear throughout this section), the task forces participated in four individual sessions during the three days of the conference.

It was the responsibility of the seven task forces to look at the original report's recommendations, study the program activity in the intervening years, and decide what really has been accomplished and what remains to be done. In other words, the participants were to answer the question, 'Where do we go from here?"

The seven task force reports that follow in this section constitute the answers to that question.

The section begins with summaries of the presentations by Lou Amabili, Clyde A. Bragdon, Jr., and Dr. John Granito on "America Burning - The Past, Present and Future," followed by the seven task force reports.

America Burning -
The Past

**Louis J. Amabili
Director
Delaware State Fire School**

The history of *America Burning* actually began in 1966 when an ad hoc group of fire service leaders met in Racine, Wisconsin, to take a long, hard look at the fire service. That meeting, known as Wingspread I, identified several problems and issues, among them the need to professionalize the fire service and the need for government to support fire service efforts.

That meeting, along with the deaths of three astronauts in a flash fire, generated enough interest in the fire problem so that the Fire Research and Safety Act of 1968 was made law. The act was in two parts. It called for a fire research program at the National Bureau of Standards (now the National Institute of Standards and Technology) and created the National Commission on Fire Prevention and Control which lead to *America Burning*.

Part one was funded; part two was not. More than two years went by and fire service leaders, somewhat concerned, met in Williamsburg, Virginia, to establish the Joint Council of National Fire Service Organizations. The council was organized in September 1970 and, for the first time, the fire service had a unified voice.

Louis J. Amabili

In November 1970, two months after the council was organized, the President appointed the members of the national commission which was funded and began work early the following year.

The commission gathered in June of 1971 in Washington, D.C., for its organizational meeting. The legislation creating the commission said it would conduct a two-year study of the nation's fire problem and report to the President of the United States.

At that meeting, we learned that it was common practice for a commission to hire someone else to do the work and produce the report. This group decided that it was going to do its own study. The work by the commission members averaged approximately a week per month for the nearly two years we were active. The commission traveled the country to hold meetings and hearings to receive broad-based exposure and input. The commission itself authored the final report, with the assistance of a professional staff and writer.

The report, *America Burning,* was released on May 4, 1973, and quickly became a classic document. It represented the expertise, concerns, comments and investments of time of 18 people who cared about the American fire problem.

The report consisted of 90 recommendations. If I had to summarize them, with the priorities established by the commission, that summary would be "prevention, detection and suppression, in that order." The report said that more emphasis must be placed on prevention. It did not say, as some would like to believe, that prevention is a panacea and suppression forces can be done away with. What the report said is that if prevention gets better, the need for suppression gets less, and that is the balance that should be sought.

The report also called for a federal or national focus on the fire problem. The U.S. Fire Administration, then the National Fire Prevention and Control Administration (NFPCA), was to be the focus for the American public, while the National Fire Academy was to be the focus for the fire service.

The report predicted a five percent annual reduction in fire deaths if the programs were funded to a level of $125 million a year. Under that funding, master planning would have received $30 million; training, $30 million; research,

$26 million; equipment, $15 million; public education, $9.6 million; fire fighter equipment, $4 million; fire data, $3.74 million; National Fire Academy, $4 million; and U.S. Fire Administration, $2.5 million. On October 19, 1974, Congress passed the National Fire Prevention and Control Act and it was signed into law three days later by President Gerald Ford.

The original organization of the NFPCA called for an administrator that was a level four Presidential appointee and a deputy administrator that was appointed by the President at level five. The administrator reported to the Secretary of Commerce.

The NFPCA had four divisions - research and development, the National Fire Data Center, National Academy for Fire Prevention and Control, and Office of Public Education.

The objective of the public education office was to overcome public apathy toward the fire problem by teaching people fire safety through public relations campaigns, publications and demonstrations. Part of the objective was conducting research to determine the most effective methods for reaching the public.

The purpose of the fire academy was to advance the professional development of fire officers and others in the fire service.

The research office was designed to handle research into fire technology, master planning, and fire codes and standards.

The data center had as its purpose a more active nationwide analysis of the fire problem that would assist in setting priorities, identifying major issue areas, monitoring fire losses and determining solutions.

The legislation also called for the establishment of the Center for Fire Research at the National Bureau of Standards (now the National Institute of Standards and Technology), with pass-through funding going to the fire administration so there would be a coordination of research activities.

While the commission had called for funding of $125 million a year, the program started off with only $6 million in 1975, jumped to $11 million in 1976 and peaked at approximately $20 million.

The acting administrator of the NFPCA was Dr. Joseph Clark. He was replaced by the agency's first permanent administrator, Howard Tipton, who had been executive director of the commission. The deputy administrator was David Lucht and the first academy superintendent was David McCormack.

This leadership put 'together a five-year plan (1976-1981) that followed the intent of the commission and the legislation. The plan was strong enough to help the federal fire programs weather the storms that they were going to run into in the future.

Those storms started with Presidential Reorganization Plan Number 3 under the Carter administration. The plan called for the integration of the federal fire programs into the newly formed Federal Emergency Management Agency (FEMA). The Joint Council of National Fire Service Organizations initially opposed this switch from the commerce department to FEMA.

However, after receiving assurances and commitments from the Carter transition team of a continued federal focus on the fire problem, the joint council rethought its position and supported the transfer. The joint council also made other concessions, including the loss of the deputy administrator's position and the inclusion of the fire programs budget into the overall FEMA budget. In retrospect, perhaps the joint council support for the move and the concessions was an error. In any case, in 1979, the USFA and NFA became part of FEMA. Later that same year, the NFA campus in Emmitsburg, Maryland, was dedicated.

Some months later, the civil defense staff college in Battle Creek, Michigan, closed down, was moved to the Emmitsburg campus and renamed the Emergency Management Institute. Now when you travel to Emmitsburg, you see the U.S. Fire Administration and National Fire Academy buried under what is now the National Emergency Training Center.

My personal concern as a former member of the national commission is that both agencies have lost identity and visibility. When you drive up Route 15 in Maryland, you see a big sign saying National Emergency Training Center. What does that mean to the average person? It has no relationship to the fire administration, the fire academy or anything else connected with fire!

As the Reagan administration came into power, members of the joint council met again with a representative of the transition team who called the fire programs an appendage to FEMA. We should have seen that as an omen because an appendage is something that sticks out like a tumor and needs to be cut off. With the proposed budget for 1982, the Reagan administration began several unsuccessful attempts to eliminate funding for, and thus dismantle, the U.S. Fire Administration, National Fire Academy and Center for Fire Research.

Those attempts were unsuccessful because the joint council and the entire fire service got involved and asked Congress to restore funding for the fire programs. While those funds were restored, a slow, but steady, erosion took place. When the attempts to zero fund the fire programs began, the budget was approximately $10 million with 40 staff members. We have ended up with a budget of approximately $4 million and 20 staff members.

To compound the problems, the fire administration has had 11 administrators, four of them permanent and seven acting, while the academy has had nine superintendents, four permanent and five acting. Every time the leadership shifted, there was a slight shift in focus that certainly added to the confusion.

Even with these problems, the federal fire programs have had significant accomplishments.

• When *America Burning* was published, you could count the number of smoke detectors in homes by the thousands. Today, we are talking about tens of millions of smoke detectors around the country.

• As a result of Project FIRES, there has been a vast improvement in the safety afforded by fire fighter protective clothing.

• In my opinion, if nothing else has come from the fire administration, funding for the development of quick-acting residential sprinklers has been a monumental achievement. Were it not for this program, residential sprinklers would not have been developed for many, many more years.

• In 1975, when the federal fire programs started, 8,800 people died in fires. Moving at the commission's goal of a five percent reduction per year, even though annual funding was short of the recommended $125 million, 14,121 people should have been saved from 1975 to 1983. In fact, 13,755 lives were saved which is pretty much on target.

Despite all the obstacles and roadblocks, things are going pretty well. America is still burning, but maybe it is not burning quite as badly as it was 15 years ago.

America Burning - The Present

Clyde A. Bragdon, Jr.
Aministrator
U.S. Fire Administration

My role is to give you an update on the status of this nation's fire problem. As you know, approximately one year ago, the 97 deaths at the DuPont Plaza Hotel fire in San Juan, Puerto Rico, generated headline after headline. I do not think there is a single issue in this country that gets more media attention than fire, when you consider all the local coverage as well.

In terms of fire deaths, we seem to have reached a plateau. The statistics are stable now and further reduction is not happening. However, everything is not measured in fire deaths. Unfortunately, they are the most dramatic of all the data we can collect.

On the other hand, arson is not stable and may even be increasing slightly.

Fire injuries present another interesting problem. I have never found two sources of data that are uniform. They go anywhere from 30,000 to 300,000 per year. Perhaps that is one of the weaknesses that can be addressed in your task force sessions. Should we have a more meaningful fire injury reporting system for this nation? We now do not really have significant data on the extent of injuries from fire.

Clyde A. Bragdon, Jr.

Let me talk about the circumstances ushered in by the Reagan era. I am a Reagan appointee, but I will try to be candid. As an earlier speaker said, we were called a negative appendage on the objectives of the Federal Emergency Management Agency (FEMA). The problem arose because there was no one on the transition team that was either familiar with, or interested in, the federal fire initiatives or the nation's fire problem.

Having had that label attached to us during those early years, we were destined to be the target of budget elimination attempts. It was the fire service and its interest groups, through their liaison with Congress, that have maintained at least a no-growth program.

Where are the federal fire programs? These programs are designed only to support those who have the basic and inherent responsibility for the delivery of fire protection, namely state and local governments. The role of the federal fire programs is to enhance the capacity of state and local governments to provide this protection. The fire programs are serving as a clearinghouse, a catalyst, a coordinator if you will, at the federal level.

I was asked at my confirmation hearings more than three years ago how the fire administration could accomplish its programs with the staffing it has. My answer is that the 1.2 million members of the American fire service would be the staff, so it was only necessary to have a skeletal federal effort.

Where do we go from here? We cannot dwell on the past. We have achievement despite adversity, adversity on the political side, and yes, adversity even within the fire service family itself at times. We need to continue to progress in the areas of prevention and suppression that were outlined in the Federal Fire Prevention and Control Act of 1974, namely engineering, enforcement, education and extinguishment.

In 1988, budget-cutting efforts took a new turn when the attempt was made to eliminate our fire prevention programs, thus taking the "fire prevention" out of the fire prevention and control act. This would have left us with only a small data operation and fire fighter health and safety programs.

The fire data program has grown significantly over the years. It now involves approximately 40 states with 10,000 fire departments. That number is growing and should peak at approximately 13,000 to 14,000 fire departments sometime during this fiscal year. This program processes more than one million fire reports annually and is an excellent example of the partnership that has been established by the federal, state and local governments.

We now have real-time fire data that interfaces with the National Fire Protection Association's (NFPA) data collection system which can do the statistical projections for those who do not contribute. Why data? This is the information age and we use data to set directions, make decisions and assist with the establishing of priorities.

The data system is basically voluntary and returns to the state and local governments data that is helpful in their decisionmaking processes. The data system currently costs approximately $350,000 a year and it is providing significant, meaningful real-time data that is used, among other things, to enhance and complement the consensus code process.

We have seen a fire fighter death rate in this country which is not acceptable. I have heard many say we should reduce civilian deaths, fire fighter deaths and fire losses to an acceptable level. I would like anyone in this room to stand up and tell me what that acceptable level might be. There is no such thing as an acceptable level. It is a goal and an objective, probably not attainable, but certainly one that can be reduced significantly.

We have seen a reduction in fire fighter deaths and injuries, a loss that costs local governments substantial amounts of money when you consider worker compensation, medical expenses, training and replacement expenses. And this does not even consider the personal costs to individuals. Much more needs to be done in this area.

Each year, in cooperation with the NFPA, we provide the "Fire Fighter Death and Injury Study." A close analysis of those studies discloses that many of the deaths and injuries are not necessary. It should not be a badge of honor to be in a profession that has one of the largest death and injury rates of any occupational group in the country. It is a disgrace and, at the same time, a challenge to move forward and eliminate these kinds of statistics. Again, that should be part of your charge.

In another area, one that cannot be emphasized enough, fire prevention was, is and will be a top priority. Fire prevention can be accomplished in several ways. A serious educational effort is probably the cheapest investment that we can make at all levels of government and in the private sector to make the public more aware of the hazards of fire. And there are a lot of exemplary programs that are doing just that. The NFPA, Tobacco Institute and other organizations have been involved in these programs.

The fire administration has programs like the 'National Awareness Campaign" which has been quite successful in terms of the amount of media attention given to it this past year.

Earlier speakers have alluded to the fact that 80% of our homes are estimated to have smoke detectors. The only problem is that up to 40% of those are non-functional. Obviously that needs to be addressed and it can be done best through the educational process.

If tire safety education efforts fail, then obviously we need some back-up systems, namely detection and early warning. We could spend this entire meeting talking about sprinkler systems and new technology, technology transfer, and the new and different applications we have witnessed recently. However, we do have a demonstrated track record of making the transfer to life safety applications with the sprinkler that was designed 100 years ago for property protection.

The data we have collected on fire deaths has given us our target audiences: the very young and the elderly. I think particularly of those more than 60 years of age who are a target because of several different reasons: the kind of housing they live in or are committed to, their economic status, their social status or the fact that they often are neglected. There are a lot of strong, interesting, positive programs at the local level addressing this problem, but the elderly are still a significant target audience.

Fire is just one of the many problems that must be dealt with by the local community as part of its social environment. However, it is unacceptable to find death from fire being relegated to the back pages of the community newspaper when some other kinds of tragedies are getting more attention.

Our National Community Volunteer Fire Prevention Program will launch right into our second National Awareness Campaign. There are a lot of resources in the local communities. It was never the federal government's intent to start funding local fire programs. It was our intent to give them seed money, a kick-off so to speak, so they would be able to generate their own independence and funding. We are probably up to 70 projects right now in 29 states and Washington, D.C. Not all of them are successful, but several certainly are.

There is much to be said just by presence and participation. That is why I like to call it a partnership. It is not the $20,000 or $30,000 that we give to the communities that is important; it is the fact that the federal government cares and would like to involve itself as a partner, but would move away once the program is able to sustain itself. That is a true partnership arrangement; that shows interest.

If nothing else, that kind of involvement is symbolic. And I really feel that symbolism has been the rallying point for the federal fire programs in recent years. The fire service had something symbolic in Washington, a symbol around which to rally by writing their representatives and senators in support of the fire programs. Many in the fire service have not attended the fire academy and have not

participated in fire administration programs, but those agencies still belong to them and they are not inclined to let somebody take them away.

There is a whole litany of items I could recite here, programs that we are involved in and have been involved in at the federal level. However, I would rather challenge you in this meeting to give us direction. And "us" does not mean the federal fire programs, it means the fire service as a whole. This symposium is not intended to be a blueprint for the federal fire programs.

This meeting was not intended to be self-serving. It was intended to address the nation's fire problem through the year 2000. By intent, the invites to this conference are a mix of people from various disciplines and organizations which affect and influence the fire problem in this country.

Certainly we would not even be here if it were not for the strong support of the nation's fire service and its interest groups. However, we cannot be continually talking to each other, so reach out, embrace and start talking to those other groups that are in attendance at this meeting. They can make a difference!

America Burning - The Future

Dr. John Granito

Dr. John Granito

The first *America Burning* group faced a tremendous game of catch-up. There had never been anything, resembling a national fire focus of that magnitude. In fact, there had been very little in the way of a national focus on the fire problem in the years preceding those efforts in the late 1960s and early 1970s. As a consequence, the recommendations of the National Commission on Fire Prevention and Control span a rather broad spectrum, ranging from suggested improvements in fire equipment and the placarding of hazardous materials transport vehicles, to the very creation of a federal agency which would put its spotlight on fire protection.

Well, several of those commission recommendations were never enacted, while others received no financing or nourishment as perhaps they should have. These observations are obvious to anybody who looks at the fire scene in the United States.

I became a fire fighter in October 1949 on my 18th birthday and, despite my presence in the fire service, you will be surprised to learn that not much progress was made over the next 20 years.

However, in the last 15 to 20 years, I think there has been tremendous progress. Some of that progress is the result of the efforts of people at the national level. But, simply put, fire protection in the U.S. has improved substantially.

For example, we know the cure for structure fires, namely, detection systems, automatic extinguishing systems, environmental monitoring systems and so forth. We understand the components of comprehensive fire prevention. In those communities where comprehensive fire prevention programs are being implemented, there have been significant improvements. In fact, the number of suppression runs in some cities has declined considerably.

Yet within this pleasant landscape, there are some terrible and terrifying scenes. Smaller communities suffer disproportionate fire and life losses. A number of metropolitan communities are struggling mightily to provide even the most minimal type of emergency medical services to their citizens. In many rural parts of the country, emergency medical service is almost non-existent.

Many, many fire departments still allocate less than one percent of their resources to fire prevention efforts. For each news article that makes us heroes (us meaning you in the fire service), at least one other, and sometimes several, indicts us, often justifiably.

So, while we have come a great distance, there remains a great challenge before us. That challenge will be met because the bywords of the fire service always have been innovation and flexibility, and I think they will serve us well as we meet the future.

I envision the future as a steamroller; it is coming at us and there obviously is not any way to stop it. We can try to run away; we can stand there, perhaps turning our back, and be run over and knocked flat; or we can meet the future

head-on. That, of course, is the purpose of a meeting such as this - to get our ducks in order, so to speak, to rejuvenate our efforts, to motivate ourselves and each other so we can meet this challenge of the future.

In thinking about the future of the fire service, most of us tend to envision a scene in some future time, after the changes have occurred and everything new is in place. We like to project ourselves to a point where a total package is all set and operating. In fact, of course, the future arrives in bits and pieces, each tomorrow following each tomorrow, and the attention is most often on the journey rather than the final destination.

What is happening in the United States that might give us some hints about the future? There are a number of concepts which are having a major impact and which portend what tomorrow and the following tomorrows will bring.

Risk management, calculated risk, trade-offs and decisions made by broad-based community groups, not just the professionals, are concepts that have taken hold in our country. People are willing to purchase some amount of protection, but they are not willing, or not able, to purchase all of the protection that the professionals specify is needed in a given instance.

Another concept is local and regional planning, with fire protection as but one element of an integrated emergency management system. There has been criticism of master planning as too cumbersome because it involves too many people, but master planning is working in many communities. There is a need to emphasize mitigation and recovery as a part of emergency management, as well as planning and response.

Someone whispered to me very recently, just a few minutes ago in fact, 'When will communities allocate the resources that need to be put in fire prevention?" I think my response is, 'When we can diminish the sense of excitement that one gets from red trucks, sirens and red lights." Fire prevention as a way to reduce fire loss is a concept that has taken hold throughout the nation.

Fire protection leadership at and from the national level, something that we did not have 20 years ago, certainly has made a significant difference.

Without any question, research, development and improved databases have been of substantial assistance to the fire protection community. In the same vein, computerization, enhanced management information systems and the increased use of non-traditional personnel in all aspects of fire protection work have helped. By non-traditional personnel, I mean civilians performing tasks that historically have been done by uniformed members of the fire service. In some communities, this is such a shocker that you dare not mention it on the first day of your visit.

These are national trends which reflect the growing concerns 'of our society. These trends and concepts are shaping our future in the fire service, as is the need for accountability and cost effectiveness.

People want more bang for the buck today, so agencies must consider adding related services to the municipal delivery package. There should be increased working relationships among agencies. There are hazardous materials incidents today that bring to the scene representatives from more than 30 agencies, each one justifiably saying, "I have a right to be here, I have responsibility in this kind of incident, and I have some authority in its command and control, and you had better listen to me."

There need to be shifts in organizational structures and staffing arrangements. We cannot deliver new services using the same structures.

We must use technology to improve efficiency and effectiveness. Those concerned with fire protection have not been able to take full advantage of technological advancements. In fact, on balance, there are some days on which I think that high technology has brought us more hazards than advantages. That

"U.S. Representative Curt Weldon (R-PA)

...Our ultimate goal...is to begin to form a caucus, a loose-knit organization of members of Congress who are willing to give the fire service more than lip service. And not just to deal with the issues at election time, but all year around. Not just to introduce bills that they know have no chance of passing, just to satisfy the constituents back home..., but members of Congress who really have a genuine interest in looking at your needs so that the next 10 years can be more productive legislatively than the last 10 years."

should not be the case. We should be taking better advantage of technology.

But still the focus is on suppression!

Now, in which directions are we moving? We certainly are moving toward automatic detection and extinguishing systems and systems that will monitor the environment, give an instant read-out and call for the necessary assistance. We are moving toward a focus on codes, a focus on proper construction, rather than just code enforcement, as part of what I call comprehensive fire prevention. And comprehensive fire prevention must include public education.

Where does this move us? When you consider that almost all our marbles have been put historically into fire suppression, one is struck with an interesting fact. Nearly everything associated with the fire service is there because we extinguish most fires with a chemical, water, transported to within approximately 30 feet of the flames by portable tubing we call hose, and advanced by humans we call fire fighters. I said it that way because I want you to think about it a bit. Suppose that we could extinguish fires in some other way that was practical and cost effective? Suppose we could extinguish fires, as indeed we can, by the generation of sound waves of a certain frequency, magnitude and amplitude, or by laser, as we certainly can? Suppose we could do this in a practical way? Think of the impact that would have on the fire suppression service, indeed I suppose, on fire prevention as well. The impact would be staggering and it is coming.

I read an article recently about the Fort Wayne (Indiana) Fire Department which is purchasing small fire trucks and using a form of dry powder on many fires, and seems to be having reasonable success with it.

If this trend toward reduced suppression, increased EMS and increased demand for additional service as a measure of cost effectiveness continues, one could predict that we would not have fire departments, but environmental protection departments. Think of the kinds of services we could provide. For example, with the growing concern for radon in our homes, is this not an area in which an environmental protection department (I realize it needs a better name) could work?

There is no doubt that fire departments are getting involved in EMS and all aspects of disaster planning and response, including natural and man-made disasters and hazardous materials incidents. But there are some new aspects as well, for example, a change in the vehicles we use to respond to an incident. Imagine, if you will, a multi-purpose department where crews are trained to operate along a broad spectrum of calls and the apparatus are called "rapid intervention vehicles," and where hazardous materials transport vehicles are tracked just like freight cars are tracked in a railroad yard.

We could tell what is coming into our communities, we could track the hazardous material and get any kind of information we needed instantly from the computer read-out. I use this as a simple application of a technology which is already with us.

Environmental monitoring of structures through cable television and similar devices is being done now in a number of communities throughout the country.

I predict that within the next two decades, the names on the sides of the vehicles will change again. They used to say fire department, and now, in many instances, they say fire and rescue department. I predict they will change to reflect multi-purpose emergency organizations designed, trained and equipped to respond to a wide variety of emergencies not covered by the law enforcement agency in the community.

The concern that I have in looking at the future has to do with the issue of who will control the nature and shape of what we currently call fire departments. Will we have structures, shapes, responsibilities and budgets forced on us by people who are not professionals in the field? Or, will we as professionals be able to shape that future ourselves? I submit to you that we will do a better job of meeting future challenges if we can shape our own futures or at least play a major role in that task.

We will play a less important role if the future is pushed on us by whomever may be pushing at the moment. I would like to feel that, no matter what happens, some future will be pressed on us by communities and citizens that need our efforts. However, I am fearful that a less important future will be thrust upon us and we will end up in a world where suppression runs are unusual.

Suppression departments will remain in existence, but they will be small organizations, with few resources, doing a job about which the public does not think or care. I hope that is not our future. I think that the fire service has much more to offer and I join with others who have been saying that the participants in this conference have a splendid opportunity to put together some recommendations which could shape the future in a way that makes sense.

In closing, what we need is a win/win situation, a situation in which our communities and citizens and their property and future are better off for our presence, a situation in which we as professionals, represented by a wide variety of organizations, are better off as well.

Task Force 1
The Nation's Fire Problem

--------- *Task Force Members:* ---------

Deputy Chief Steven C. Bailey,
Seattle (WA) Fire Department,
National Fire Information Council

Dr. Richard E. Bland,
Pennsylvania State University

Mr. Peter Brigham, American
Burn Association

Mr. John C. Gerard, National Fire
Protection Association

Mr. John Glenn Hart, III, U.S.
Fire Administration

Fire Commissioner Charles A. Henry,
State of Pennsylvania

Mr. Matt Kane, National League of
Cities

Mr. William Randleman, Fire Chief
Magazine

Mr. Fred S. Ringler, People's
Firehouse, Inc.

Mr. Philip Schaenman, Tri Data
Corporation

Ms Barbara Lundquist, Tri Data Corporation, Facilitator

I. Introduction

The objectives of Task Force 1 were to identify new issues, problems and trends associated with the overall fire protection problem in the United States. Upon a review of Chapter 1, "The Nation's Fire Problem," and Chapter 2, "Living Victims of the Tragedy," the task force members were to update the state of the nation's fire problem as defined in *America Burning*. In addition, the task force was assigned the job of developing recommendations that will guide future fire protection analysis and planning.

II. Background

As a result of a series of hearings and related research, the National Commission on Fire Prevention and Control prepared a narrative description of the fire problem as perceived in 1972 - 1973. Many of the recommendations made by the commission to alleviate these problems have been initiated. For example, the commission recommended that Congress establish a U.S. Fire Administration (USFA) to provide a national focus for the nation's fire problem and to promote a comprehensive program with

caid, and support of medical research and training have contributed to this progress. In addition, the specialized training required for burn treatment personnel is being carried out on a broad scale by professional associations, emergency medical service programs and the burn centers themselves.

How has the fire problem changed since 1973? The number of fire deaths has decreased by approximately 25% and the number of fires reported to the fire department has decreased by 20%. Nevertheless, the United States, along with Canada, still has the worst fire death rate for all the industrialized countries for which we have comparable data. The U.S. fire deaths per million population are almost twice the average fire death rates for other industrialized countries.

Per fire, our loss (in deaths or in dollars) is not high. However, the number of fires in the U.S., even figured on a per capita basis, is extremely high. This suggests that our fire suppression efforts are relatively good, and that to reduce fire deaths and dollar losses we need to concentrate on fire prevention.

adequate funding to reduce life and property loss from fire. The U.S. Fire Administration was indeed established in 1975, along with the National Fire Academy and the Center for Fire Research. The annual budget of the federal fire programs, however, has been consistently much lower than the funding levels recommended by the commission.

The commission also was concerned with upgrading the nation's burn treatment facilities. While there are no specific programs to support this recommendation, much of it has been accomplished. By 1983, there were approximately 125 burn centers in the U.S., including one in virtually every metropolitan area. Such federal programs as Medicare and Medi-

Over the past five years, we have averaged more than 6,000 dead annually, more than 100,000 injured, with two and a half million fires reported to the fire service. Many, many more fires go unreported each year, causing additional personal injuries and property loss.

The total cost of fire in the United States, including losses plus the cost of protection (built-in systems, fire department and insurance overhead), is $30 billion a year. This does not include the value of labor donated by more than a million volunteer fire fighters or such indirect fire costs as medical expenses.

With respect to deaths, young children (under 5) have a much higher risk of being killed in a fire than teenagers and adults -50% to 100% higher than the average population. The elderly (over 65) are at highest risk - triple that of the rest of the population. They sense fires less easily, move less quickly and are killed more easily by burns and smoke than younger people.

In addition to age, there is a racial and ethnic dynamic to the nation's fire-loss. Blacks have a fire death rate almost double the national average. American Indians have an even higher death rate than blacks. Differences among these and other groups may result from differences in family income, education levels, family stability, alcoholism levels, and other social practices and problems which correlate with fire incidence and fire death rates.

Since *America Burning*, the responsibilities of fire departments across this country have expanded significantly to include most of the delivery of emergency medical services in our communities, and the management of both hazardous materials incidents and natural disasters. Some estimate that emergency medical responsibility alone has more than doubled the demands for manpower, training, and management time and attention in our fire departments.

America Burning did lead us to develop a federal focus, a set of key agencies, namely, the USFA, National Fire Academy and Center for Fire Research. A National Fire Data System has been established in which 40 states are participating. Interest and action on the part of the fire service in the field of public fire education has increased significantly. A

national cadre of thousands of fire fighters, trained at the National Fire Academy in technical areas, prevention and management, is moving into and up the officer ranks in local tire departments. And the absolute number of fire deaths occurring each year has gone down.

But a document cannot provide leadership. The federal fire programs are not comprehensive. They have no continuity and they are trivial in size as federal programs go. Inadequate staffing is as serious a problem as inadequate funding. Even the National Fire Data System is fragile and may falter without more support. Today, federal staffing to manage a data system that collects detailed information on almost a million fires a year is down to two people.

The need remains to analyze and communicate what we have. Much of the discussion of Task Force 1 concerning the "problems" underlying the nation's fire problem centered around critical information gaps and the need for more timely and effective dissemination of the information we have. This information is needed both for planning effective fire prevention programs and raising the awareness of the seriousness of the fire problem with the public, elected officials and even the fire service itself.

III. Critical Issues

The theme that ran throughout the discussion among Task Force 1 members about the nation's fire problem was ",.Let's build on our successes to meet tomorrow's challenges." There was early consensus that the recommendations of *America Burning* are still valid, and that they will continue to be in the coming years. With *America Burning* as a master plan, many successful programs have been delivered through federal, state and local governments. Based on our successful track record of having a fire focal point to offer technical and financial assistance "when available," the task force members felt that the effort to control and reduce the fire problem in the United States will continue to be successful.

The following is a discussion of the key issues identified by Task Force 1. These are not ranked in priority.

1) Lack of awareness. We have failed to convince elected officials of the seriousness of the fire problem. While the fire service has been effective in lobbying for specific fire protection-related state laws and local ordinances, its effectiveness at the federal level has been limited. By and large, lobbying efforts have been ineffective when they were targeted at gaining support for the program budgets and staff of the federal fire agencies.

The U.S. Congress (as well as other elected and appointed officials at the state and local levels) needs, first and foremost, to be educated about the concerns of the fire service, the seriousness of fire and other emergency management problems, and the methods available to address these problems. Another failure in the area of lobbying is that the fire service has not worked with such other parties as community service organizations and industry with whom they might productively form partnerships to bring greater influence to bear on public officials. Unless this problem is solved, adequate resources will never become available.

2) The need to understand and target the underlying causes of fire and fire losses. The underlying causes discussed range from major national problems, such as poverty, to cultural orientation, meaning the absence of any kind of punishment or stigma when a fire occurs.

One precept of this discussion was that much of the public education effort currently undertaken by the fire service reaches middle-class or upper middle-class children and adults in mid-size communities and suburban areas. The fire departments in the best position to undertake public education serve an audience less in need of it than other communities. Another way to look at this, however, is that even in America's "safer" communities, the risk of dying in a fire is as great or almost as great as it is in most other countries overall. Our worst areas are much, much worse than the fire risks faced in other countries, but even our best communities are bad.

The specific underlying causes about which more information is needed and which, at this time, people believe should be targeted directly by public fire safety education are:

- poverty;
- inadequate housing;
- public apathy;
- lack of knowledge about fire;
- lack of stigma when you have a fire;
- lack of knowledge that there is a fire problem;
- alcohol and drug abuse;
- lack of adult supervision of children; and
- lack of qualified supervision of other dependent populations.

3) Information needs. We need to fill information gaps within our body of knowledge about the nation's fire problem and to extend that body of knowledge to include EMS and hazardous materials requirements. Some of the specific information gaps we need to fill include:

- reason for fire problem changes;

- profiles of burn victims, their demographics, severity, and major causes of fires that result in serious burn injuries;

- the total count of injuries from fire overall;

- time to flashover - a field measure of this is needed;

- EMS - magnitude and characteristics of the problem; and

- hazardous materials - magnitude and characteristics of the problem.

4) Defining the fire problem. The role of the fire service in America has expanded

greatly. Responsibilities have been broadened by public expectation, public mandate, and the interest and concern growing within the fire service itself. EMS and hazardous materials are the two major new areas of demand. Fire service resources in contrast have not grown at a comparable rate. Personnel, money and equipment/materials have been diffused across these major areas. Fire prevention, particularly public fire education, has suffered because of this.

6) The maintenance of the positive impact of smoke detectors and its extension to high-fire-risk population groups. The installation of smoke detectors in homes has been a landmark change in the fire environment. Smoke detectors are believed to have been the most significant factor in the decline of fire deaths in America. However, people with the highest fire rates have the fewest detectors, and today, maintenance problems are eroding detector effectiveness. The smoke detector ranks along with only two or three other devices - the centrifugal pump, combustion engine and the radio - as technologies that have, in their time, advanced significantly the effectiveness of fire protection delivery.

IV. Task Force Recommendations

Many ways of attacking the above problems were discussed. The specific solutions that Task Force 1 recommended as most important to undertake as soon as possible are listed below.

1) Failing to convince elected officials of the seriousness of the fire death, injury and loss statistics was considered the most serious problem because it is the path to resolving many other problems. The task force felt that success in this area would open the door to resources needed for the other problems. A number of recommended solutions target this first problem. Below, the most important alternate solutions are discussed first, followed by other solutions considered in the general discussion of how to solve this overall problem:

- fire service organizations need to find their common ground and present that with one voice;

- we need to support and educate the new Congressional Fire Services Caucus;

- we must brutalize the fire problem to make it more vivid to the public and to officials;

- the base of organizations having input and/or membership in the Joint Council of National Fire Service Organizations should be broadened; and

- locally, the fire service should establish citizens committees or involve community groups to garner support for fire prevention programs.

A simple tool, lacking today, that would help in undertaking any or all of these actions is a brief, accurate description of the nation's fire problem. Seven to eight years ago, when the federal fire programs were getting considerably more support from Congress and the White House, this was available. It came in the form of *Highlights of Fire in the United States*, the U.S. fire death map and charts showing bar graphs of population groups at risk and the fire death trend line. Why are these no longer available?

It is believed that the majority of public officials at all levels of government rank the fire problem very low as a topic of concern to them. It is disproportionately low in their concern, particularly when its seriousness is compared to such other problems as crime or housing. In parallel, even officials who are concerned are not convinced how the fire problem can be solved. Some do not believe public fire education works. Unfortunately, there is little documentation of successful local programs that can be used to convince them. A few case studies of program effectiveness in the field of fire prevention would be a valuable tool in this area.

Within the fire service, concerned organizations have pulled in different ways instead of pulling together. Leadership in fire is fragmented. Some "fragments" are self-serving. Officials, both legislative and executive branch members, do not know to whom to listen. Further, important groups are excluded.

The burn/medical community, community service groups and others are not included on the joint council. A broader base is needed. It is believed this would have the effect of helping the joint council achieve agreement on a few broad principles and programs to support, and would help it get beyond its image of being constituted of a number of separate organizations continually at odds with each other, no one of which is large or strong enough to be significant, particularly when compared to the interest groups lobbying in other areas of national concern.

At the local level, cities that have established fire-related citizens committees appear to educate their local officials more effectively. This model should be promoted in cities across the country as a mechanism for significantly increasing resources available in fire prevention. It also is known that cities having good fire data available appear to be more effective at educating their officials.

Nationally, fire-concerned organizations have not lobbied effectively. A particular problem has been that they have failed to follow up. Learning how to lobby should be a priority for any fire community leader seeking state or national office in a fire service organization. Lobbying support should be a major aspect of fire association staff work.

2) We need to understand and target the underlying causes of our fire problem. To solve this, we know what is needed first - money. The USFA already has the legislative mandate to undertake these studies, but lacks the money and manpower to carry out such a research program. The USFA should conduct or sponsor this research, if and when funds are made

available. The research findings, as they become available, can be used as "news" in public education campaigns to raise the awareness and concern of targeted fire risk audiences, as well as public officials and others concerned with fire protection.

3) The solution to the problem of filling major information gaps in the nation's fire problem is clear - money and manpower. To address the problem of filling major information gaps in the description of the nation's fire problem, some of the preliminary discussion focused on the fact that there is a lot of good data now available. It is in use in many fire departments and in some states. But greater data analysis capability needs to be developed throughout the fire service, and resources are not available for analyzing the data now collected and for doing special studies. From outside the fire area, we need to capture and publicize available information on the burn injury problem and to integrate burn and fire data collection.

Some specific recommendations are that the USFA should support the preparation of model data analysis reports from different size departments for other departments to copy. The National Fire Academy's executive development courses should improve data analysis training. Companion information on successful ways to tackle a major fire problem should be identified and disseminated, along with the statistics regarding that problem, for example, the fire problems of the elderly and inner-city juveniles.

Consistent support is needed to keep the national fire incident reporting system going. Analyzed data no longer is getting back to local fire departments. Feedback is essential both to encourage

alties at unreported fires that similarly are not reported to anyone. Information on the seriousness of burn injuries is one of the most dramatic tools for educating and motivating elected officials and the public. We do know that burns are the most severe fire injuries, that the ratio between males and females is 2 to 1, and that the overall severity of burns has gone down. According to experts in the burn treatment field, out of the 6,000 fire deaths each year, 2,000 die in burn centers. But beyond this, little is known.

their participation and to improve the service they provide. More USFA staffing is needed here, as well as increased funding.

In the area where major gaps of information exist, for example, why there have been changes in the trends of the major causes of fire, funding and staffing are needed to oversee the research. Soft scientific research is needed on the human behavior and attitude aspects of the fire problem. There is a study underway by the USFA on the demographics of fire fatalities. This study is expected to reveal some new information on high-fire-risk population groups, but to date, no money has been available to examine the causes of fire and successful approaches to preventing it.

The added problem in this area is that little is known about burn victims. There is virtually no integration of fire and burn injury data. All casualties at reported fires are not necessarily reported anywhere. In addition, there are many casu-

The treatment of burn victims should be included as part of our concern for the overall fire problem. The federal Health Care Finance Administration should develop a payment system sensitive to the special problems of burn centers. Government and relevant professional groups need to address the increasing nursing shortage problem which, in turn, is affecting burn treatment. Further, federal support of research in burn treatment and rehabilitation programs is needed.

4) *To a certain extent, the need to "redefine the fire problem" was addressed in America Burning.* It does contain recommendations in the areas of EMS and hazardous materials incidents and other emergency response problems. However, the federal agencies established under the Fire Prevention and Control Act of 1974 are concerned almost exclusively

with fire. They do not have either the clear mandate or the resources to serve as a federal focus in these other areas, regardless of their importance to the fire service and the public. Task Force 1 felt that adequate funding, attention and staffing at the federal level was needed, not additional legislation. Everyone agreed that EMS and hazardous materials responsibilities are here to stay. Some say fire departments are being drained of fire protection resources because they perform these other functions. Alternately, some say fire departments have their current resources only because of these other functions.

An enhancement to the federal programs, either as part of the existing agencies or as special federal task forces on these issues, should be considered.

Someone needs to communicate to the fire world what is going on at the federal level and, in turn, to represent the interests of the fire service to Congress and the regulatory agencies. Congress needs to be more sensitive to the impact of their actions on state and local government. The passage of the Superfund legislation has been a classic example of an instance where the USFA should have been out front communicating the status and implications of the legislation to the fire community and representing the interests of the fire service.There is not a single fire association that has the resources to monitor the actions of the federal government and Congress regularly and to keep the fire service adequately informed. The USFA needs the resources to provide this information clearinghouse. It is critical both to the fire service and to the policymakers in Washington.

5) *Smoke detectors are no longer new.* There have been many campaigns promoting detectors over the years, and, as a result, they are used widely. We know, however, that the households having fires do not have detectors, and that many detectors are maintained poorly. The use of life safety systems in homes is probably one of the most significant changes in the fire environment, but lack of maintenance is believed to be cutting down on smoke detector effectiveness. To achieve greater reduction in fire deaths, the increased use and maintenance of smoke detectors is required. Many in the fire service believe that changing the behavior of lower income families is much harder than changing the behavior of "middle Americans."

Two primary solutions to this problem were recommended. One is additional federal resources to identify and publicize successful local maintenance programs. The other is a solution that also was recommended with regard to attacking some of the underlying causes of fire and influencing elected officials. Namely, the task force recommended establishing citizens committees or involving community groups to a much greater extent than is done today to secure the resources and involvement of enough people at the local level to implement effective programs.

Supplemental Report to Task Force 1

prepared by
Peter Brigham, MSW, and
Arnold Luterman, MD
American Burn Association

When *America Burning* was written in 1973, its authors were handicapped by a serious shortage of data on burn injury and other aspects of the fire problem. There was little understanding at the time of the distribution and severity of burn injury in the general population. No general consensus existed as to which burns should be treated in general hospitals and which in specialized treatment centers. While there was considerable concern about the potential financial impact of burn centers on their parent institutions, there was little experience to rely on because the initial burn centers, for the most part, were located in specially funded research facilities. With these factors in view, *America Burning* expressed serious concern as to the sufficiency of burn treatment facilities and staffing in the United States.

America Burning also recommended special training of those providing emergency care for burn victims in general hospitals. This training is being carried out now on a broad scale by professional organizations, emergency medical service programs and burn centers themselves at the regional level throughout the country.

By focusing the interest of the fire service on the burn injury problem, *America Burning* also helped to forge local alliances between burn center hospitals and the fire service which continue to this day. Fire service leaders in many cities have played important roles in fundraising for the initial development and ongoing operation of burn centers. Their interest was often crucial in convincing hospital administrators that a decision to develop a burn center would receive broad public support. In the early days, fire service leaders provided credible and valuable independent testimony on the importance of burn centers. As burn centers became established, the focus of the burn center/fire service alliance at both the local and national level broadened to include efforts to prevent burn injury.

Burn centers currently admit approximately 20,000 patients per year. The average length of stay per patient is 15 days. More than 90% of these patients survive and are discharged to resume their lives outside the hospital. Approximately one-third of surviving burn center patients will return to the hospital at least once for additional surgery. Another 50,000 burn patients with less severe injuries are admitted to all other hospitals. At least 10 times that number are treated and discharged from hospital emergency departments.

Admissions to burn centers held steady through the mid-1980s, while burn admissions to other hospitals declined. This trend reflects at least two phenomena. Physicians and administrators in general hospitals have increased their referrals of burn patients to burn centers, and the total number of serious burn injuries has declined. At the same time, pre-hospital patient care and transportation, often carried out under fire service auspices, have improved, contributing to the smooth functioning of the burn care system in most regions.

In *America Burning*, recommendations also were made for the funding of research in burn injury and smoke inhalation. In the mid-1970s, the National Institute of General Medical Sciences granted funds for burn research programs at seven burn center hospitals; other public grants supported research at other burn centers.

The dissemination of research findings and the increased clinical experience resulting from the concentration of severe burn patients in burn centers has resulted in improved survival rates and improved functioning of survivors following these injuries. According to statistics compiled by the National Burn Information Exchange, the "LA 50" of hospitalized burn patients (the size injury at which half of all patients will survive) has increased in most age groups to more than 60% body surface area. This is approximately double the size injury producing 50% mortality 20 years ago. While there is no comparable data on improvements in quality of life for survivors, there have been great advances in surgical and therapeutic techniques during the same period.

Clearly, there has been considerable progress over the past 15 years in the areas where recommendations were made in *America Burning* - burn center development, training and staffing for burn treatment in burn centers and general hospitals, and research.

The current challenge is to sustain the burn care system which has developed in the past 15-20 years, while continuing to improve the survival and quality of life outcomes of burn victims. Major areas of concern and related recommendations for federal action are listed below.

1. Burn Center Reimbursement

Many burn centers are facing increasing financial problems. This has resulted from growing numbers of indigent patients, and changing payment methods which tend to discriminate against burn centers. This problem can be expected to worsen, as burn center survival rates continue to improve, and as additional payer sources adopt the new reimbursement methods.

Recommendation The federal Health Care Finance Administration should review the entire financial situation facing hospitals with burn centers.

2. Burn Center Staffing

The nation is facing a dramatically increasing shortage of nurses. This has an especially serious impact on burn centers, with their high staff-to-patient ratios and the requirement for a high level of nursing skill. The total number of hospital-oriented nurses must be increased substantially.

Recommendation: The Congress and the Department of Health and Human Services must work with the appropriate professional groups in addressing this problem.

3. Burn Research Funding

Major problems remain in burn treatment. These require a strengthened federal commitment to burn research. Improved survival from burn injury has produced a growing number of patients with residual disabilities. A concerted research effort must be initiated to improve our ability to rehabilitate this group of patients.

Recommendation: The National Institutes of Health should increase significantly their sponsorship of research in burn treatment and rehabilitation.

Task Force 2
The Fire Services: Operations

─────────── Task Force Members: ───────────

Mr. Tom Brennan, Fire Engineering Magazine

Chief Warren E. Isman, Fairfax County (VA) Fire and Rescue Department, International Association of Fire Chiefs

Chief Roger McGary, Takoma Park (MO) Fire Department, International Society of Fire Service Instructors

Chief Edward J. Phipps, San Francisco (CA) Fire Department

Mr. Wayne Sandford, Connecticut Commission on Fire Prevention and Control, National Association of State Directors of Fire Training and Education

Mr. William M. Neville, Jr., National Fire Academy, Facilitator

I. Introduction

How can fire protection be improved?" The easiest answer to this question is to increase fire department budgets to buy more equipment and hire more staff. This alternative was rejected immediately by the National Commission on Fire Prevention and Control not only because significantly higher budgets were unlikely, but also because increased funding does not necessarily produce increased effectiveness. A more practical alternative lay in reviewing fire service priorities and modes of operation in order to increase service delivery effectiveness and efficiency.

The objectives of Task Force 2 were to assess the operational requirements and possibilities for current and future fire service activities and responsibilities. This assessment occurred within the context of the issues and recommendations discussed in Chapters 3-7 of *America Burning*.

II. Background

While basic fire service operations essentially have remained unchanged for more than 50 years, change within the last decade has been significant and wide ranging. This metamorphosis undoubtedly will continue to unfold well into the next century. For example, many fire departments have revised operational priorities to meet demands for emergency medical services, hazardous materials incidents and non-fire suppression responses. Other factors, such as reduced funding, the Equal Employment Opportunity Act, Age Discrimination in Employment Act, Fair Labor Standards Act and Superfund reauthorization, have changed the way fire departments recruit, train and manage personnel.

In 1973, personnel assigned to suppression companies generally would spend the majority of their active time training and responding to fire alarms. Now, structural fires have decreased, and *80%* of the alarms may be for emergency medical incidents. The remainder of a fire fighter's time is devoted to public education, code enforcement inspections, community relations and training/physical fitness activities.

The labor pool for both volunteer and career departments has changed as well. On the positive side, many in the fire service have developed a more professional stance. Increased demands for greater technical and managerial skills and more career opportunities have been contributing factors. For volunteer departments, this change poses significant challenges because the increase in training, education and responsibilities has impeded the recruitment and retention of personnel. In addition, inadequately prepared staff can result in legal liability and greater public safety risk.

Chapters *3-7* of *America Burning* included recommendations to place a higher priority on fire prevention activities, the recruitment of women, increasing multi-jurisdictional cooperation, the

use of fire protection master planning, service delivery and personnel productivity improvements, advanced and specialized education and training, emergency medical services, improvements in fire fighter protective equipment, equipment standardization, and the establishment of the National Fire Academy.

These topics may or may not remain critical issues. However, the contemporary critical issues might include current and projected missions and functions of the fire service, alternative processes and techniques for fire protection, shift schedules under the Fair Labor Standards Act, fire fighter health and safety, physical fitness, company staffing, in-service activities, training, the Incident Command System, inspection/enforcement, public education, community relations, reflex time criteria for station location and staffing levels, apparatus size and design, burnout, alcohol and drug abuse, and communications and dispatching.

III. Critical Issues

The task force members felt that the future goals and values of the nation's fire service must meet a two-fold test. They still must be responsive to the traditional, although diminishing, functions of fire suppression, while simultaneously becoming more dynamic to meet the challenges of a rapidly advancing technologi-

---

U.S. Representative Sherwood P. Boehlert (R-NY)

"What America Burning concluded in 1973 remains a valid agenda today, yet the products of our past successes are threatened and we are still building the coalitions needed to achieve new triumphs."

cal world. The reordering of priorities and redirection cannot come from external sources. The fire service needs to make its own decisions and be in control of its own fate.

The following describes the five critical issues identified by the task force. These are organized in priority of importance.

1) Human resource development. There is a real need for the fire service to develop its own human resources in order to provide imaginative leadership and decisionmaking capability with a local, national and international perspective. The fire service leadership must be given the tools to develop these resources,

The task force identified four key components of this issue:

a. upgrading officer selection and training which must start at the level of the company officer;

b. recruiting, training and retaining volunteer fire fighters;

c. recruiting minorities and women; and,

d. training allied professionals.

2) What is the mission of the fire service? The task force felt that the fire service's goals and objectives are too narrow to meet society's changing needs. This problem is compounded by the public's perception of the fire service's mission, which is often inaccurate. The confusion then frequently results in resistance to change at all levels. Many fire service officers instinctively reject new equipment, procedures and duties. The ongoing debate over company fire safety inspections, greater priority for fire prevention, and expanded emergency medical services are examples of this tension.

The task force felt that resistance to change also occurs when society thrusts change and new responsibilities on the fire service. Further, society often requires that the fire service perform these additional tasks and functions without the accompanying resources and training.

3) The lack of modern technology and methods. The fire service needs national standards to replace the Insurance Services Organization Grading Schedule. New standards and methodologies are needed to guide staffing requirements, the deployment of equipment and apparatus, and the location of stations. Of course, new and innovative technologies have been developed within the past decade, including expanded automatic and mutual aid agreements, the Integrated Emergency Management System, civilians in fire suppression and training positions, and master planning. However, additional work is required to integrate these systems into the fire service and increase their credibility.

A continuing problem experienced in the fire service is the lack of information for choosing and evaluating equipment and apparatus. A "consumers guide" would be very helpful.

The inability to plan on a comprehensive basis is limiting effectiveness and efficiency. Some tools, for example, master planning and the FIRESCOPE products, are available, but more need to be developed.

4) "Instant" implementation of affirmative action adversely affects Standards and the overall quality of personnel. Because all personnel have the same responsibility, they all should be trained to the same standard, and double standards must be eliminated. This problem

is part of a much larger and fundamental issue, namely, the lack of standardized minimum personnel requirements throughout the fire service.

Often, the lack of national standards forces the local departments to develop their own. This places the departments in a difficult and hazardous position. Unless these requirements are validated by national experience, departments may use biased and inaccurate requirements which will affect the quality of the personnel and leave the departments open to discrimination complaints.

Such national professional qualification standards as the National Fire Protection Association 1000 series do provide a benchmark, but any standard will need to be validated and used nationally throughout the fire service. Whether the personnel are career or volunteer fire fighters is not a valid consideration for defining this issue. "'A fire behaves the same way no matter who has to deal with

it." The task force raised the possibility that it may be impossible to train volunteers for all the functions expected of the career fire fighter. The impact of federal Occupational Safety and Health Administration training requirements for hazardous materials responders is an example of the dilemma. A double standard between volunteer and career personnel poses legal liability and public safety risks.

5) *Federal, state and local government intervention.* The task force members felt that governments on all levels were passing along a lot of responsibilities to the fire service without adequate funding or training. In addition, demands for "instant compliance" with these imposed responsibilities lower morale and standards,

In addition to the critical issues above, the task force reviewed the following problem areas (these are not ranked in priority):

- lack of alternate funding sources;
- failure of managers, mayors, public and media to appreciate the positive impact of emergency services or understand procedures;
- lack of fire protection training for construction engineers and architects;
- failure of the fire service to become involved with other municipal managers;
- lack of time to schedule required training;
- lack of standard pre-training for minorities and women;
- lack of resources to deal with chemical abuse;
- obsolete standard operating procedures (SOPs) or lack of emergency sops;
- inconsistent and unreliable fire loss data;
- lack of information on new technology failures and successes;

- lack of standardization in communications and dispatching systems - problem for mutual or automatic aid;
- failure of tie service management to become involved in national or regional fire service issues; and
- lack of professionalism at all levels of the fire service.

IV. Task Force Recommendations

The task force members chose to develop solution strategies for the highest priority issue - development of human resources.

The task force recommended a comprehensive program of research conducted by the U.S. Fire Administration to develop minimum standards for all fire service personnel. These standards then would be implemented by the National Fire Academy through training and certification programs. Specific program areas include:

1. *developing a job analysis of fire service positions (volunteer and career) on a broad nationwide basis;*

2. *determining the applicability of existing fire service position (volunteer and career) professional qualifications (NFPA 1000);*

3. *determining appropriate selection procedures for fire service positions (volunteer and career);*

4. *examining volunteer membership trends;*

5. *researching the validity of pre-employment training for women and minorities;*

6. *researching the credibility of the fire service with allied professionals, and elected and appointed oficials; and*

7. *researching the impact of mandatory and voluntary certification programs on volunteer and career personnel.*

Once this research is conducted, the National Fire Academy should develop a program to promote and coordinate the research results with the state and local agencies responsible for training and certifying public safety personnel. Engineering and architectural organizations should be involved in this process as well.

The task force developed a three-phase implementation strategy for this national employment research and development program.

Phase 1 - Short Term: The fire service should participate actively in extra-jurisdictional issues (e.g, surveys, committees) and national fire service organizations.

Phase 2 - Medium Term: The fire service should budget to attend national standards meetings and training sessions (e.g., NFA, NFPA, ASTM), should adopt the national recruitment and employment development program, and should support performance-based selection processes (both volunteer and career).

Phase 3 - Long Term: The fire service should adopt validated training programs for minorities and women, should have a comprehensive officer development program, and should formalize the evaluation of state, metro, and county training programs.

Task Force 3
The Fire Services:
Management and Administration

———————— *Task Force Members:* ————————

Chief Paul Boecker, Lisle-Woodridge (IL) Fire Department, International Fire Service Training Association

Commissioner Joseph Bruno, New York City (NY) Fire Department

Mr. William H. Hansell, Jr., International City Management Association

Mr. Tom Herz, International Association of Fire Fighters

Chief Jerry Knight, St. Petersburg (FL) Fire Department

Mr. Joseph O'Hagan, U.S. Army and Federal Fire Service Task Force

Chief John B. Stewart, Jr., Hartford (CT) Fire Department

Mr. Clarence Williams, International Association of Black Professional Fire Fighters

Mr. Garry L. Briese, International Association of Fire Chiefs, Facilitator

I. Introduction

Task Force 3's area of concern lay in the issues and recommendations discussed in Chapters 3 through 7 of *America Burning*. The objectives of the task force were to review the key managerial and administrative issues in the current and future fire protection environment, identify the requirements and constraints associated with these issues, and develop alternate strategies for meeting the requirements.

II. Background

Fire chiefs and other fire service management and administrative personnel have faced a wide range of major changes over the 15 years since the publication of *America Burning*. These changes have included expanded departmental responsibilities, reduced operating budgets, increased personal and departmental legal liability, revised personnel management practices resulting from the Fair Labor Standards and Equal Employ-

the only nationally recognized document which specifies resource requirements as a function of community size and hazards.

The fire protection master planning process was intended to provide a basis for determining departmental requirements. Even this approach, however, used selected components of the ISO *Grading Schedule* because there are no other quantitative standards for making such determinations as the required number and location of stations and companies, and corresponding staffing levels.

Many of the other concerns raised by the National Commission on Fire Prevention and Control have been addressed. For example, significant efforts have been made to bring women and disadvantaged minorities into the fire services. To attenuate existing bias, affirmative action programs have been instituted, and actions have been initiated to reexamine and revise recruitment assessment guides and promotion examinations. The status and priority of fire prevention has undergone a major change. Many fire service leaders see code enforcement, public education and arson control as core elements in combatting the fire problem. More significantly, fire prevention positions are being viewed as key rungs on the career ladder in many departments. The National Fire Academy is established and provides educational and training

ment Opportunity Acts, heightened employee health and safety concerns, and increased competition for fewer resources.

In many cases, chief fire officers depended on the Insurance Services Office (ISO) *Grading Schedule* as the basis for fire protection planning and the rationale for justifying personnel, equipment and facilities. However, the influence and value of this insurance-based planning guide have been declining as local government questioned its applicability and the cost-effectiveness of its requirements. In addition, the ISO has stated that the *Grading Schedule* was prepared for insurance industry use, and not as a fire department planning manual. Thus, fire chiefs generally have lost

U.S. Representative Sherwood P. Boehlert (R-NY)

"There is little question that the work of defending the existing fire programs will have to continue, and we hope that this conference, by focusing attention on the seriousness of the issue, will make the coming battles a little bit easier."

opportunities where none existed previously. Finally, the federal fire programs have conducted research to increase the effectiveness of fire fighter protective equipment.

III. Critical Issues

The task force participants emphasized that the problems facing today's fire service must be faced by both career and volunteer fire managers. The problems must be recognized by large and small, urban and rural, and career and volunteer fire departments.

A theme that ran throughout the many hours of discussion was that the fire service frequently has the authority and resources to solve many of its own problems. This theme is repeated at all levels of the fire service.

The problems that face today's fire service manager are primarily "people problems."

Many of the issues can be handled internally by the fire service. Continued support and financial encouragement are needed (a true partnership) from the U.S. Fire Administration to facilitate these changes.

The task force members felt that the "America Burning Revisited" conference must reinforce the positive accomplishments of the original *America Burning*, otherwise the report of this conference will not be complete. The many accomplishments of the past 15 years should be acknowledged and highlighted. However, many of the same problems discussed by the commission in 1973 exist today, and some of these have grown worse over the

years. In addition, the past 15 years have presented the fire service with many new problems that were not anticipated in the original report. Therefore, the list of problems and action strategies should be even longer than that in the original report.

The future of the fire service is being decided from outside by such groups as public managers, insurance companies and the federal government. It is clearly not in the best interests of the fire service to allow this to happen, and the fire service should work with these outside special interest groups to gain more influence over the external forces that define its mission and environment.

cers, and their continuing education will be one of the primary challenges facing the fire service in the 1980s and 1990s. Existing fire chiefs, managers and administrators must continue with their management education as well.

Labor and management, career and volunteer fire fighters, fire and non-fire organizations must quickly begin to work together in order to regain control of the future.

This conference must serve as a call to those concerned with the future of fire and emergency services to renew and redouble their efforts. The conference must serve to refocus the attention of the fire service community on the absolute necessity of working together and recognizing that we can set our own future course.

The task force identified five key problem areas. Ranked in priority, they are:

* fire and emergency service health and safety;

* no adequate method for the evaluation of fire and emergency service organizations and operations;

* personnel management;

* planning; and

* the need for a shift of emphasis from suppression to prevention.

The task force members identified numerous other issues that impede the cost-effective delivery of fire protection services. These auxiliary concerns were either included within the scope of the top five issues or were not considered a high priority. The following issues are not ranked in priority.

This outside world views fire differently from those in the fire service. While fire may not be seen as a major problem by non-fire service citizens, government managers and politicians, fire departments frequently are viewed as a problem. The fire service is not "selling" its programs and activities, nor is the fire service taking "credit" for what has been accomplished. Further, these outside groups are demanding change faster than the majority of fire service agencies can make, or are willing to make, the changes.

We must recognize that the future of the fire service is with its younger offi-

1. Compensation programs for employees have no relationship to efforts to prevent emergency incidents or losses.

2. The fire service needs more accurate information on the fire problem; all jurisdictions (100% of all fire departments) should be participating in the fire incident reporting system and in management information systems.

3. The public should not regard fire as socially acceptable.

4. A local, regional and /or national equipment deployment model should be developed.

5. National effectiveness measures and standards are needed to help give fire managers benchmarks to evaluate performance.

6. Arson prevention and detection measures should be upgraded.

7. There is a need for fire chiefs to lead, or at least participate in, an integrated emergency management system in their areas.

8. There is a need for improved training at all levels, including chief fire executive.

9. There is a need to improve fire-police relationships.

10. The firehouse should be used as a community base or resource.

U.S. Representative Doug Walgren (D-PA)

"...The thing that stands out when you look at America Burning is that we clearly have failed to implement it..."

11. There is inadequate use of civilian staff in the fire service.

12. All fire fighter safety and health issues should be given a higher priority.

13. Fire departments must deal effectively with hazardous materials (right-to-know legislation).

14. Continuing education for chief fire executives should not be ignored.

15. The fire service must discuss career/ volunteer personnel issues.

16. There is a lack of certification standards.

17. Support for master planning has been discontinued.

18. There is a lack of innovative revenue sources to relieve taxes or allow for expansion of services.

19. There is a lack of fire-safe design and materials in the working and living environments.

20. The fire service is ineffective politically at the local, state and federal levels.

21. Federal/state mandated actions that add responsibility without additional finding should not be tolerated by the fire service.

22. Programs must be initiated to prevent /reduce fire deaths and injuries in poorer communities.

23. Fire and building officials do not communicate about, nor lobby for, fire and life safety issues.

24. *Unions must be brought on bo ard with changing management philosophies.*

25. *Collective bargaining agreements limit the potential of management and employees for increasing productivity and initiating innovation.*

26. *AIDS and communicable disease issues must be addressed*

27. *There is inadequate recruitment and promotion of women, blacks and other minorities (equal employment opportunity).*

28. *Hazardous material right -to-know issues must be addressed.*

29. *Fire departments must be staffed and equipped (including apparatus) adequately.*

30. *Fire chiefs must change with the times.*

31. *No adequate measuring standards or instruments exist for fire department evaluation.*

32. *The roles of fire chief as department manager and head fire fighter should be examined.*

33. *Ways should be studied to organize local government so it can make better policies and deliver services designed to prevent and meet emergency situations.*

34. *Greater emphasis should be placed on fire prevention to allow for rapid implementation at local level.*

35. *The fire service must deal with substance abuse, abuse prevention and employee assistance program (EAP) issues.*

36. *Information about innovations must be shared in order to increase the effectiveness of fire prot ection service delivery,*

37. *There are too many fire fighter injuries and deaths, and not enough research into fire fighter safety and health.*

38. *A shift of emphasis from suppression to fire prevention and education must be accomplished without increased resources.*

V. Task Force Recommendations

The recommendations of the task force are organized around the five most important issues for the management and administration of the fire service today and in the future. Both the problem areas and the solution strategies identified to address these problems are listed by priority.

1) Fire and emergency service health and safety.

a   Programs should be implemented to provide for insurance loss reimbursements to governments or fire departments for fire fighter injuries or deaths on the fireground.

b. National Fire Protection Association Standard 1500 should be recognized as a guideline by all fire departments.

c. A program should be developed to test and rate fire equipment and apparatus. The task force members felt that a system similar to the Underwriters Laboratory rating program would be a good model for this type of program.

d. More sophisticated and in-depth data on health and safety issues should be collected and disseminated.

e. The U.S. Fire Administration should take a leadership role in hazardous materials issues at the federal level.

f. Duplication of right-to-know legislation should be avoided in state and federal legislation.

g. Physical fitness standards should be developed, adapted, validated and enforced for all fire and emergency service personnel.

h. AIDS is a major concern to emergency medical service personnel. This illness is a communicable disease and should be treated as a public health issue, not a civil rights issue.

Federal legislation is needed to educate emergency service personnel concerning their exposure to AIDS and other communicable diseases.

i. Research programs on fire and emergency service health and safety issues should be re-initiated. These research programs should include, but not be limited to, physiological aspects, psychological aspects, toxicity concerns and protective equipment.

j. Employee assistance programs (EAPs) should be designed and made available for fire and emergency personnel and their families.

k. Any programs imposed on the fire service by federal or state governments should be accompanied by adequate funding to pay for the program, or the mechanism to provide such funding.

2) No adequate method exists for the evaluation of fire and emergency service organizations and operations.

a. Establish a national evaluation method to measure the effectiveness and efficiency of fire and emergency service agencies and systems. The development of this methodology should involve a broad spectrum of fire and emergency service, governmental management and non-governmental resources.

b. Chief fire officers should recognize, lead and implement the Integrated Emergency Management System (IEMS) concepts into fire and emergency service planning and operations at the local level.

3) Personnel management.

a. Encourage the development of labor-management committees to focus attention on productivity and innovation (given the constraints of labor agreements or legislation).

b. Encourage aggressive implementation of equal employment opportunity programs, involving recruitment, retention and promotion of minority groups in the fire and emergency services for both career and volunteer departments.

c. Develop and distribute specific programs and publications to assist volunteer fire departments in recruiting and retaining volunteer fire fighters.

d. Management education should be targeted at all position levels within the fire service. Educational opportunities should be made available at the local, state and national levels, as well as at the National Fire Academy.

e. There should be an identification of the skills and knowledge necessary for the position of fire chief (volunteer or career, large or small, urban or rural). Continuing education requirements and the certification of chief fire officers should be considered.

f. Management courses taught at the National Fire Academy should be evaluated and updated every two years by a team of active fire chiefs and public managers (for example, members of the International City Management Association).

4) Planning.

a. Master planning must be reinstated as a major program of the U.S. Fire Administration and National Fire Academy.

b. Fire departments must use the concepts of master planning in the development of long-range goals and objectives.

c. The U.S. Fire Administration should initiate the evaluation of alternative delivery systems for fire and emergency services.

d. Local communities must involve the fire service in all building and fire code development, including its adoption and enforcement.

e. Any mandated actions by the federal or state governments must be accompanied by adequate funding or funding mechanisms (fiscal impact).

f. Research should be done on the identification of alternative revenue sources for fire and emergency services, with a special emphasis placed on user fees to better distribute costs between the public and private sectors.

g. Research should be encouraged to identify fire-safe designs in materials.

h. Fire departments must participate in community-based programs and should use fire stations as the focus of the program.

5) Shift the emphasis of the fire department from suppression to prevention.

No solution strategies were identified for this final issue because of a lack of time.

Task Force 4
Fire and the Built Environment

--- Task Force Members: ---

Mr. George B. Barney, Portland Cement Association

Mr. Paul Fitzgerald, Factory Mutual Corporation

Mr. Earl Flanagan, U.S. Department of Housing and Urban Development

Mr. Rolf Jensen, P.E., Society of Fire Protection Engineers

Ms Janet Kimmerly, Firehouse Magazine

Mr. Richard P. Kuchnicki, Council of American Building Officials

Mr. Peter G. Sparber, The Tobacco Institute

Mr. Henry Roux, Armstrong World Industries, Inc., Facilitator

I. Introduction

The objectives of Task Force 4 were to define the fire protection environment vis-a-vis the built environment, assess the progress that has been made since the publication of *America Burning* and recommend strategies for improving building fire safety.

II. Background

The fire protection aspects of the built environment have changed dramatically during this century. Building and fire codes have eliminated many hazards, for example, inadequate exits, and have required such new fire safety features as smoke detectors and sprinklers. However, new building designs (e.g., atriums, high-rises), construction and furnishing materials, and uses (e.g., the processing of hazardous materials) have created new hazards and have dramatically increased safety risks. Futhermore, it is expected that this process of eliminating and creating fire safety hazards will continue in the future. For example, future demands for affordable housing may result in increased densities and pressure for the use of lower-cost materials that may compromise fire safety. Trade-offs of current fire code requirements in return for installation of automatic detection and suppression systems may become a common approach to providing fire protection in the built environment.

Significant changes in the building and fire codes often have occurred as a result of major fires. The MGM Grand

for materials, building fire safety design, realistic material test standards and procedures, the systems approach to fire safety, fire safety effectiveness statements, consumer product design, fire safety education for architects and engineers, community code adoption and enforcement, use of smoke detectors, hazardous materials, and motor vehicle, aircraft, marine and railroad fire safety.

A lot of progress has been made since 1973. The Center for Fire Research has been a leader in researching the basic processes of ignition and combustion. It has provided technical assistance to the Consumer Product Safety Commission's efforts to establish mandatory and voluntary flammability standards in such products as mattresses, upholstered furniture, heating equipment and cigarettes. CFR has developed methods for predicting the growth and spread of fire and has conducted a broad program of research into suppression and detection technologies.

The fire community has gained some control of many fire variables. Fire departments conduct surveys of commercial and institutional buildings, inspecting for fire hazards and compliance with fire codes. Cigarette ignitions of bedding and upholstered furniture used to be a major cause of fires and accompanying deaths and injuries. The mattress flammability standard promulgated by the Consumer Product Safety Commission, and the voluntary Upholstered Furniture Action Council (UFAC) flammability program has significantly improved this hazard area.

Additional fire standards developed by the Center for Fire Research and American Society for Testing and Materials (ASTM), as well as additional mattress

Hotel fire in 1980 resulted in revisions in Nevada, and influenced other states and communities to review safety requirements for hotels and motels. Of course, it would be preferable that a comprehensive, long-range program for achieving fire safety in the built environment be established and implemented prior to disasters that result in death and injury.

Chapters 8-12 of *America Burning* included the following topics: residential fire safety, commercial and industrial fire protection, major fires, building materials and furnishings, toxic products, inadequate building codes, flammable fabrics, occupancy fire loading, fireworks, combustion dynamics, combustion standards and upholstered furniture regulations required by such jurisdictions as Boston and California, have resulted in even safer products. Other building product tests, for example, the pill test for drapes and carpets and the National Institute of Building Sciences' combustion toxicity test effort, will result in additional life safety.

Code compliance has always been a problem. But the code enforcement process has changed substantially within the past 15 years. The majority of states have adopted model building and fire codes as minimum building requirements. Many local jurisdictions supplement these state requirements by adopting their own codes. Fire departments have expanded their fire prevention programs, and state and local governments have established training requirements for personnel involved in inspection and enforcement activities.

IV. Critical Issues

The task force's discussion on fire safety in the built environment focused on problems related to public behavior, technology and code enforcement. The range of topics that were raised underscores the complexity of this issue. The five key concerns are organized in priority.

1) Lack of public awareness of the fire problem. The public does not recognize or realize the consequences of a fire. A 1976 U.S. Fire Administration study of public attitudes with regard to residential fire safety determined that the fire safety issue was not a prominent one. New information obtained by the USFA confirms this conclusion. Focus groups held

around the country show that people are barely aware that there is a fire problem, much less aware of its scope and seriousness.

2) Lack of control of the building content. There is a need for combustion and toxicity data on products used to construct and furnish buildings. The problem of not being able to control the contents of a building is aggravated further when its occupancy or use changes, and the contents then may represent an even greater danger.

3) Fire hazard methodology. The task force noted that there are really no adequate methods for conducting a fire hazard analysis, The engineering tools needed to undertake this task are not presently available. Some research has been conducted in this area by the Center for Fire Research, and the Hazard I methodology provides a critical starting point. The Hazard I analysis protocol is applicable to single-family dwellings and permits the calculation or modeling of fire growth by tracking the development and distribution of heat, smoke and toxic gases. The objective of the model is to be able to predict the probability of occupants being able to escape from the dwelling during a fire.

4) Fire safety education. The task force lamented the poor training or lack of training of the many actors involved in the construction and code enforcement system. This lapse ran the gamut from fire fighters to more senior code enforcement officials to architects and engineers. Fire fighters need to have more knowledge about building construction, not only to conduct code enforcement inspection, but to combat fires (e.g., to anticipate potential roof collapse on the fireground). It is most unfortunate that fire safety is not given a higher priority in architectural and engineering curricula.

There is not a cohesive education for fire protection engineers, starting from a bachelor's degree up to a doctorate. The standards and certification criteria needed to coordinate this effort are just not available. Further, there are too few academic programs, and none for the advanced student. The

Illinois Institute of Technology's fire science program was discontinued in 1984, as was Harvard's doctoral program. A masters degree is offered at Worcester Polytechnic Institute and only a bachelors degree at the University of Maryland. Further, while there are a number of certificate and associate degree programs offered by the model code organizations, community colleges, associations and the National Fire Academy, their programs lack the required comprehensiveness and uniformity.

Much of this problem stems from a lack of leadership within the fire community, as well as the awareness concerns expressed in Issue 1 of this task force. Many in the code and design professions are unaware of the need to design fire-safe buildings. There is a reluctance even on the part of some fire officials to accept the recommendations of the Society of Fire Protection Engineers.

The model code arena is another area where design input from fire safety officials and engineers is limited.

Often it has been members of the fire protection community that have taken a back seat to initiatives that were of prime concern to them. For example, the American Society of Civil Engineers is now in the process of adopting a national fire protection design standard for the performance of structures. The process was started without input from the SFPE. While the SFPE is involved now and is partici-

> *"U.S. Representative Curt Weldon (R-PA)*
>
> *The purpose of this (caucus) is not to dictate to you what should be done, but rather to listen to the needs of the fire service, take information from gatherings such as this, from meetings that are held at the National Fire Academy or around the country, and then to provide the legislative supportpolitically to help you turn those recommendations into actions."*

pating in the deliberations of other model code organizations, the example is illustrative of the need for the fire community to take a leadership role in areas with which it is concerned.

5) The problem of existing buildings. The majority of structures that will be used for residential, commercial and industrial occupancies in the year 2000 are already built. Many of these properties have hazards and risks that have not been addressed by current building codes. What's to be done? Buildings can be retrofitted with sprinkler and detection systems and other fire protection measures. With some notable exceptions (e.g., hotels/motels in Florida and Nevada, and high-rises in New York), few state and local building codes require existing buildings to meet current code requirements, unless the building is being renovated substantially. Some building owners and managers, in particular the Marriott Corporation, are voluntarily retrofitting their buildings with sprinkler systems. However, they are the exception, not the rule.

Other issues discussed by the task force, but not considered a priority, included:

- the lack of casualty fire loss data;

- the traditionally low priority of fire prevention within most fire departments;

- the cost of improved, safer building materials and furnishings;

- competitive safety issues (e.g., radon) and competitive building features;

- the cost of overprotection - redundant systems;

- the cost and quality of sprinkler and detection systems and their installation;

- the lack of national uniformity in building/fire codes;

- the lack of information on the storage of hazardous materials;

- the lack of a national insurance program;

- the disunity and combativeness among the members of the building community (i.e., owners, manufacturers, fire service, building officials);

- the lack of fire department involvement in new construction; and

- subjective fire test methods.

IV. Task Force Recommendations

The task force prepared solution strategies for the first two critical issues. The issues are ranked in priority, as are the recommended solutions.

1) Lack of awareness. Regarding the public's lack of awareness to the fire problem, the task force recommended the following solutions.

a. Develop an awareness plan that would start with a survey of the dynamics of the public's current attitude. This survey would update the 1976 USFA study which examined public attitudes with regard to residential fire safety. The survey then would be supplemented with objective case studies examining both the tangible and intangible consequences of fire. Specific incidents are important because the task force wanted to "brutalize" the fire experience in order to make it appear more vivid and realistic. The final stage of the plan would be to develop an implementation strategy outlining target audiences and methods for communicating the message to them.

b. The task force felt that the most effective overall strategy for cultivating a sensitivity to fire hazards was to start communicating this message at an early age. Fire education needs to begin with kindergarten and be reinforced continually in succeeding years. People need to understand that there are consequences to their behavior. Positive, fire-safe behavior is rewarded, while negative or careless behavior ends in the tragedy and loss resulting from fire.

c. Adults need to be educated about fire safety as well. The task force recommended a public relations program aimed at encouraging the public to be a part of the overall solution to the

fire problem. Adults need to learn what steps they can take to minimize the ignition and growth of fire. Suggested approaches include the use of the electronic and print media (e.g., movies, novels), and an effort similar to the "Neighborhood Watch" programs that have been developed successfully by many police departments to combat home burglaries and other neighborhood crimes.

d. Incentives that encourage fire safety could have a beneficial effect. Tax, insurance, legal and social incentives, as well as deterrents, would go a long way toward modifying both personal and business behavior.

2) Lack of control over the building contents. The task force approached this problem in an unusual manner. Instead of controlling the fire hazards and risks associated with construction materials and furnishings, we need to control the fire performance of the room or dwelling. If information is available on the room's

rate of heat release and total heat content and how it will perform during a fire, it is possible to calculate whether the room would go to flashover. With this information, limits could be established so that the room would not go to flashover and the fire would burn itself out. This information also would be useful for designing a sprinkler system, an approach being taken by Factory Mutual for non-residential occupancies. For residential occupancies, this information could be used to design the response pattern and location of detection systems, as well as fire barrier requirements. Instead of using general fire rating requirements derived from some consensus fire scenario, a builder could install fire walls or partitions that are tailored to the specific hazards and risks of the building's contents relative to its use. However, in order to implement this approach, two types of information are required.

a. First, research is needed to develop tests and standards in the areas ofignitability, fire growth (i.e., rates of mass loss and heat release), combustion toxicity and suppressibility.

b. Second, 'we need information on the fire characteristics of building contents. All products used in buildings should have a label describing their heat release performance, total heat content and gas generation characteristics. The publication and labelling of these should be a "requirement." There are some initiatives underway both in this country and abroad that provide some basic information on the pros and cons of fire hazard labelling. For example, West Germany has started a voluntary "right-to-know" project called the Blue Eagle Program. Under this program, manufacturers

voluntarily label their products as being environmentally safe. The manufacturers are modestly enthusiastic about the program and have found it an effective marketing tool.

Attachment 1
Other Recommended Solutions

The task force identified a number of alternative solution strategies for the first two priority issues discussed in the body of its report. These alternatives are listed by issue. The task force members did not assign a priority to these suggested solutions.

1. Lack of awareness

a. Require product manufacturers to install fire safety instructions.

b. Require an Underwriters Laboratory label on all building contents.

C. Conduct a national survey of the perception of the residential fire problem.

d. Provide a fire safety instruction sheet to all hotel guests.

e. Have meeting planners check for fire safety.

f. Have fire departments launch a fire safety program in the schools.

g. Promote Fire Prevention Week.

h. Promote Building Safety Week with the National Conference of States on Building Codes and Standards (NCSBCS) and Council of American Building Officials (CABO).

i. Continue "Learn Not To Burn" in schools and on television.

j. Require fire drills in public buildings, schools, offices and nursing
homes.

l. Bequest government subsidies of
fire safety equipment and public
service advertisements.

k. Use building wardens for fire
safety education and inspection.

m. Involve the American Bed Cross
and similar organizations in fire
safety endeavors.

n. Require builders to include fire
safety instructions in new homes.

o. Brutalize fire consequences to
make fire more vivid to the public.

p. Use the media to promote fire
safety - movies, radio, public
service announcements, newspaper articles, soap operas, etc.

q. Coordinate existing fire safety
programs.

r. Initiate a public relations campaign.

s. Create tax and insurance incentives and disincentives for fire
behavior.

t. Increase the public visibility of
local fire departments.

u. Offer free home inspections by
local fire departments.

2. *Lack of control over building contents*

a. Institute an upholstered furniture
flammability study, examining
both fire experience and possible
new regulatory requirements.

b. Increase fire department inspections of building contents - mandatory for non-residential occupan-

cies and voluntary for residential
occupancies.

c. Begin additional flammability
research on building contents for
non-residential occupancies.

d. Require product manufacturers to
develop and implement a fire
safety awareness program.

e. Base insurance rates on the hazards and risks posed by occupancy
and building contents.

f. Mandate code inspections when
any change of building occupancy
occurs - non-residential.

g. Require all product manufacturers
to publish fire growth characteristics of products stored, shipped
and used.

h. Establish a heat release limit for
rooms to prevent flashover.

i. Develop heat release criteria for
designing sprinklers.

j. Develop heat release or mass loss
information to design detector
systems and barriers.

k. Develop combustion/toxicity tests.

l. Require all building owners to
publish and label fire growth characteristics for all rooms.

m. Develop an occupancy certification
system.

n. Hold building owners legally liable
to record occupant/occupancy
changes.

o. Improve cost-effective fire protection control systems for building
contents; develop maximum standard for fuel loading.

Task Force 5
Fire and the Rural Wildlands Environment

--- Task Force Members: ---

Chief Louis J. Amabili, Delaware
State Fire School

Mr. Colin A. Campbell, Alexandria,
Virginia

Chief Mary D. Chambers, Bernalillo
County (NM) Fire District

Mr. Olin Greene, Oregon State Fire
Marshal

Chief Richard Lamb, Wayne
Township (IN) Volunteer Fire De-
partment, National Volunteer Fire
Council

Mr. James B. Roberts, Maryland
Department of Natural Resources

Mr. Kenneth H. Stewart, National
Criminal Justice Association

Mr. Gerard Hoetmer, International City
Management Association, Facilitator

I. Introduction

Task Force 6 concentrated on the
issues and recommendations discussed in
Chapters 13 and 14 of *America Burning*. The group's task was to evaluate the
issues facing rural and wildland fire
protection in light of current and future
requirements for controlling fire in these
environments. In addition, the task force
members were to identify alternate
solution strategies for meeting those
requirements.

II. Background

The majority of the nation's fire
departments protect small communities
and rural areas. These departments face
a multitude of problems, including limited resources, long response times,
inadequate (or nonexistent) water supplies, declining volunteer participation,
lack of adequate training and a dramatic
increase in hazardous materials incidents. The major problem may be the
provision of a sufficient number of adequately trained personnel who can respond on a timely basis. Many volunteer
fire departments have reported difficul-
ties in recruiting and retaining sufficient
numbers of personnel.

These personnel problems result from
a variety of reasons, including people
working outside of the community, reduced opportunity for fire fighting (without a corresponding interest in emergency
medical services), and a general apathy
concerning voluntarism. Employers may
be reluctant to permit personnel to leave

work to respond to an alarm, especially when these alarms are for emergency medical incidents rather than fires which are more visible and dramatic.

Fire protection for wildlands has been complicated by the construction of homes and other occupancies in the urbanized communities bordering the wildlands. Life safety is now a much larger issue for the wildland environment. There have been several significant developments in wildland fire protection since 1973. For example, the FIRESCOPE Program, which produced the Incident Command System, was conducted originally to improve suppression of wildland fires. Fire prevention for wildland areas has

included the reduction and modification of fuel, prescribed burning, adoption of special codes for involved structures and public education.

Arson in rural and wildland areas is of continued concern. The lack of expertise and resources to detect incendiary fires, conduct an investigation and pursue prosecution may encourage arson in these areas.

III. Critical Issues

Chapters 13 and 14 of *America Burning* included the following topics: rural residential fire safety, long response times, volunteer fire fighter training, regionalization, water supplies, small community and rural master planning, transportation fires, rural arson, private brigades, forest fire prevention, incendiary forest fires, urban-wildland interface, fuel management, weather modification, model state forest/grassland fire protection laws and their enforcement, public education, fire hazard and weather forecasting, and wildland fire suppression techniques (including air attack).

The task force identified the following seven problem areas which are organized by priority.

1) Public education. The highest priority problem identified by the task force was public education or, more accurately, the public's lack of awareness about the fire problem. The public does not seem to understand the fire hazards associated with rural areas and wildlands. Unfortunately, this lack of awareness affects key political support of, and funding for, fire protection services. A public relations/information campaign needs to be waged to turn this problem around. However, the lack of leadership and expertise will impede this effort. Basic information is unavailable. For example, there is little behavioral information available that examines the public's attitude toward the fire problem and suggests possible ways to overcome it.

2) Insufficient land use planning. The task force felt that inadequate land use planning is a critical flaw in rural development and an impediment to a comprehensive fire protection strategy. Land use and fire prevention and control master planning go hand in hand. Zoning is an important tool for controlling the fire problem. For example, growth or development patterns in many areas can be regulated by the availability of adequate water supplies for fighting fires. Land use planning can regulate access for apparatus and the density of subdivisions, all of which are important fire prevention and control variables.

The urbanization of rural and wildland areas is producing a number of new fire problems. As more people build residences in remote, poorly regulated areas, there has been an accompanying increase in building/fire code violations, improper use and storage of hazardous materials, and inadequate access roads for emergency apparatus. Recognizing this new problem, a collaborative effort among the U.S. Forest Service, U.S. Fire Administration and National Fire Protection Association was undertaken to identify the problems and search for solutions.

3) Code development. The fire service has too little input into the development of building and fire codes. For the rural and wildland areas, codes are often

inappropriate. In other words, codes designed for cities or suburban areas can have less relevancy for rural or forested areas. For instance, the task force discussed the issues of combustible building materials, particularly roofing, and fuel load clearing which have a much greater impact on the fire problem in more remote areas. A task force member cited a wildland fire during which houses were saved, in neighborhoods that were devastated, because those houses were built on a more secure location and combustible fuel around the houses had been removed.

If the fire service is going to respond to this issue, its national leaders and local executives are going to have to be more active and aggressive. The fire service typically has taken a backseat approach on codes and fire prevention. More training in code development and enforcement for chief fire officers is needed.

4) Training. During the past 15 years, much progress has been made in the area of fire service training. The National Fire Academy is among the many organizations that have had a major impact on the management and technical capabilities of fire service leaders. Nevertheless, more needs to be done. For many rural departments, state and county agency training programs are not readily or easily available. The problem, however, goes beyond access because even when these training sessions are available, volunteers have difficulty attending them because of work commitments or penalties levied by employers.

Rural arson is becoming a more substantial problem. Training resources and efforts need to be marshaled to respond to this problem.

5) Fire protection resources. Fire protection in rural and wildland areas often involves interagency cooperation, Unfortunately, even with the models provided through FIRESCOPE and IEMS, communication among agencies both within and across jurisdictions remains poor. New work needs to be done in the area of inter- and intra-jurisdiction relationships. Action plans to control major fires and conflagrations would benefit greatly from this better coordination among key actors.

Fire service organizations need resources to deliver required services. These necessary resources include funding, community commitment, equipment and personnel. The task force felt that all of these elements needed additional

support. If fire prevention and control services are going to meet the requirements set by the community, the fire service needs the tools to accomplish the job.

6) Insurance. Insurance plays a significant role in fire safety. It can provide incentives, as well as disincentives, for fire safety behavior. Unfortunately, many insurance companies misunderstand the dynamics of risk associated with fire. Underwriting practices stress loss potential and not good practice or fire safety behavior. Sprinklers are an example of this approach. While sprinklers reduce the risk of large fire loss and premium credits are given for them, overall insurance savings can be negated by increased costs for potential loss from water damage.

7) Technology. The answer to the fire problem lies in behavior and technology, Built-in protection that lowers fire risks and hazards can be realized through the development and use of less combustible and/or toxic building materials and the use of automatic fire protection systems.

The requirements and constraints of the rural and wildland environment are different from more urban areas, and this difference needs to be reflected in building and fire codes.

> *"U.S. Representative Curt Weldon (R-PA)*
>
> *. . Legislators.., have got to be held accountable for their votes, and their leadership or lack of it, that they take on your national priorities here in Washington."*

IV. Task Forces Recommendations

The task force identified solution strategies for all of the seven problem areas. The strategies are ranked in priority by both the importance of the problem area and the solutions.

1) Public education

a. The fire service must market itself and its mission, The fire service needs to make better use of the electronic and print media for these efforts. The initial step of the public relations plan is to conduct a needs assessment to determine the necessary details and scope of the education effort. Research needs to be conducted to identify target audiences, their attitudes toward fire safety and appropriate methods for communicating with them.

b. Educate the political officials to their responsibilities and roles in rural fire protection and control.

c. The national fire information exchange network needs to be reinstated so that information on successful approaches can be shared.

d. The fire service needs to communicate with other groups in the jurisdiction to promote fire safety. The fire service no longer can afford to talk to itself. Coalitions need to be established with key interests in the jurisdiction, e.g., other public officials, builders, architects, educators and business leaders.

2) Lund use planning

a. Urge greater use of fire prevention and control master planning techniques by political jurisdictions.

b. Seek low-cost loans for water conservation and supply systems through the Rural Development Act.

c. Educate local officials to the dangers and hazards in the wildland environment. Officials need to know about the dynamics of fire, the cost of fire protection and fire loss data.

d. Develop model state laws and codes relating to fire protection in the forest, e.g., burning permits, zoning and building regulations, fire safety devices on motorized equipment, construction of fire breaks, access and escape routes, and emergency measures for the closing of the woods.

3) Code development

a. Codes and standards appropriate for rural areas must be developed, adopted and enforced.

b. Publicize code enforcement successes. The public needs to know that good fire safety practices work. For example, citizens need to know that the use of non-combustible roofing material can save their house.

c. Local executives need training in code development and enforcement.

4) Training

a. State and county agencies should make training easily available to rural fire departments.

b. Legislation should be introduced at the state or national level to allow volunteers to attend training sessions or to respond to an emergency incident without penalty from employers.

c. Continue to distribute the National Fire Academy training programs.

d. Develop a training needs assessment system in order to ensure that programs concentrate on the issues and needs of the fire service.

e. The scheduling of training programs must be tailored to the unique time constraints of volunteers.

d. The stipend program at the National Fire Academy needs to be secure. Most volunteer and rural fire departments do not have the resources to participate without financial assistance.

e. Strengthen state training programs with funds and expertise.

f. Unify the different incident command systems being used around the country, communications techniques and procedures should be compatible.

g. Develop a national system for forecasting fuel build-up.

5) Fire protection resources

a. Institute programs to coordinate local and state group purchasing of equipment and supplies in order to reduce unit costs through greater economies of scale.

b. Provide added funding in Title IV of the Rural Development Act for the training and equipping of local fire protection forces.

c. Rural and wildland fire departments need to reach out to other groups in the community, state and nation to obtain their involvement and increase access to resources. The fire service needs to educate and inform elected officials of their responsibilities and roles in fire protection. The fire

service needs to communicate with other rural departments in the national wildfire coordination group. Training in community development techniques should be made available to fire departments.

The fire service must involve the community in identifying the fire problem, as well as the services and resources required to address it. National, state and local training organizations should expand theirprograms to teach these skills.

d. Provide reasonable and justifiable funding. Existing funding vehicles established through the Farmers Home

Administration and the Rural Development Act should provide no- or low-interest loans.

e. Rural master planning needs to be encouraged. Fire departments need to use this tool for organizing and delivering fire protection services effectively and efficiently. Master planning methodologies need to be developed for the wildland environment.

f. New ideas should be encouraged, developed and disseminated throughout the fire service. Such innovative ideas as using retirees for fire prevention programs, employing civilians in fire command positions, and consoli-

dating building and fire inspection services have increased the efficiency and effectiveness of many fire programs. Continued work in this area is essential.

g. Establish a national fire weather service.

h. Encourage use of the National Volunteer Fire Council's recruitment and retention programs.

i. Data is essential for managing fire protection programs. All fire departments should be contributing their fire loss data to the National Fire Incident Reporting System. The U.S. Fire Administration should develop incentives to encourage fire departments to contribute.

6) *Insurance*

a. Urge the National Association of Insurance Commissioners to seek legislation in each state to require on-site inspection of any dwelling prior to any purchase and subsequent renewal of insurance.

b. Establish a coalition of insurance and fire service interests to develop practical incentives and penalties to modify fire safety behavior. For example, there could be a premium credit or increase based on whether inhabited areas are clear of fuel load.

c. Develop a train-the-trainer fire safety course for the insurance industry and require it for licensing.

7) *Technology*

a. Increased funding should be made available for technological research. Information is needed on the fire performance characteristics of a room, fire suppression equipment, and the combustion/toxicity characteristics of building materials and furnishings.

b. Establish national standards for the combustibility of construction materials (e.g., roofing shingles). These standards could be implemented through the model building codes or NFPA 101, *The Life Safety Code.*

c. Develop a residential sprinkler standard appropriate for rural use; the standard should take into account such characteristics as longer response times and less water supply.

d. Institute a comprehensive technology transfer program that would compile and disseminate information on new techniques, studies and programs. It is essential that this information be distributed widely to local fire officials.

Task Force 6
Fire Prevention

Task Force Members:

Ms Travis Cain, Office of Juvenile Justice and Delinquency Prevention, U.S. Department of Justice

Deputy Chief Joseph M. DeMeo, New York City (NY) Fire Department

Chief Charles H., Kime, Phoenix (AZ) Fire Department

Dr. Arnold Luterman, MD, University of South Alabama Burn Center

Mr. Edward H. McCormack, International Society of Fire Service Instructors

Chief Richard Moreno, Tucson (AZ) Fire Department

Mr. Patrick Murphy, U.S. Conference of Mayors

Chief Bert T. Parker, Washington County (OR) Fire District No. 1

Dr. Anne Phillips, MD, National Smoke, Fire and Burn Institute

Mr. James V. Ryan, American Association of Retired Persons

Mr. Robert B. Smith, Fire Marshals Association of North America

Mr. Robert Burns, Fire Loss Management Systems, Facilitator

I. Introduction

Chapters 15 - 17 of *America Burning* included discussions on a wide range of fire prevention activities: fire safety education, fire safety for the home, and fire safety for the young, old and handicapped. The National Commission on Fire Prevention and Control noted that fire prevention was assigned too frequently a much lower priority than other fire department activities, particularly suppression. Further, fire service personnel avoided fire prevention assignments because many perceived, often correctly, that these positions are excluded from the fast track of fire service career paths.

The commission recognized that changing the realities and perceptions underlying fire prevention was a key element in any plan to combat the fire problem because fire is dealt with more effectively and efficiently if prevented or controlled early. The commission recommended a series of actions to increase the impact of fire prevention. The mission of Task Force 6 was to review these recommendations in light of the progress made over the past 15 years and recommend new strategies to address current and anticipated fire protection requirements.

II. Background

Fire prevention was given major emphasis in *America Burning*. The commission recognized that there were limits to the capabilities of suppression to control fire loss. In fact, the commission noted that communities which rely too heavily on suppression forces for their fire protection may be assuming extraordinarily higher risks. Moreover, the commission reflected that this approach was also inefficient; it was unlikely that increasing the resources for fire suppression (i.e., apparatus, personnel) would result in proportionate reductions in fire losses.

Fire protection consists largely of two types of activities, namely, suppression and prevention. Suppression essentially is a reaction to events. Resources are mobilized and deployed following an incident - the ignition of fire. Prevention is more active. It involves the effort to decrease the chances of fire and, failing this, to control its destruction by methods

that are independent of actions taken after ignition occurs (e.g., education, code enforcement).

Much has been accomplished since 1973. Many (if not most) of the nation's fire departments have significantly increased fire prevention activities over the past decade. In-service company fire safety inspections and participation in public education programs are now common. Fire departments have enlarged their role in supervising the fire safety of the built environment. They are more involved in overall land use planning and reviewing plans for proposed developments. Fire departments frequently are involved in the development and adoption of model fire codes and the acceptance of local amendments to strengthen these codes (e.g., codes requiring installation of smoke detectors in existing dwellings).

While conclusive data is not yet available, many experts feel that the increased emphasis on fire prevention has contributed to the declining number of reported fires and fire deaths. Fire prevention activities may become even more important in the future if fire department resources continue to decline, following the maxim, "...it is cheaper to prevent a fire than to fight it." Also, community volunteers may become key members of a fire prevention program, especially because the training and physical fitness requirements differ from those required in fire suppression,

III. Critical Issues

While Task Force 6 identified more than 20 problem areas that need to be addressed to increase the effectiveness of the nation's fire prevention efforts, it concentrated on the two most important issues. These are rank ordered by priority.

1) Lack of knowledge and recognition of the fire problem and the value of fire prevention. While much yet needs to be done, a number of programs have been developed within the last 15 years that effectively address this area of concern. Some of these public education programs provide an excellent base for continued work, including NFPA's "Learn Not To Burn" program, the National Smoke, Fire and Burn Institute's "Smoke Drills," Exit Drills in the Home (EDITH), and the Smokey Bear fire safety program.

One measure of success is the degree of concern about the fire problem on the part of many private and public organizations and the amount of effort directed at education programs.

Ironically, the sheer number of programs and organizations working in this area is part of the problem. There is a need to evaluate the effectiveness of these programs. Are they communicating the correct message? Are they targeted at the right audience? There are really no systematic answers to these questions. Moreover, while the proliferation of organizations and media promoting fire safety is a blessing, they rarely are coordinated at the national, regional or, even, local levels. This disorganization produces a mixed, confused message to the public and wastes scarce resources.

There has been a shift in the building and furnishings industries toward fire safety. Pressure from the fire service, the federal government and model code changes have produced significant gains in the last 15 years. Mattress and upholstered furniture resistance to cigarette ignition has improved greatly. Work is being conducted on cigarette fire safety. The fire service needs to continue these efforts. More incentives (and disincentives) need to be developed to reduce fire hazards.

2) Lack of support for a standardized reporting system to document needs/ problems and program results. The task force recognized that many of the problems just noted revolve around the inability to identify systematically problem areas and evaluate alternatives for addressing them. The problem comes down to the lack of data. The task force reflected that much of its work would have been easier if it had more information for analyzing problems and selecting solutions to resolve them.

The task force noted that the national fire data system is inadequate. Data is not collected in a uniform manner, nor is there a national focal point to coordinate and direct its collection. There is data coming from burn agencies, hospitals, medical facilities and the fire service. But how good is this information? How valid is it? Training and quality control programs set up originally by the U.S. Fire Administration have not been supported. Task force members raised questions about the quality of the data entered into the system.

While a system for collecting data has been established and is operated by the USFA, the information needed by the fire service is not available. Of course, there is much more information today than there was 15 years ago, but this just underscores the problem. The effort has not been a complete one. There is information that can rank order the origins of fire, geographical fire problems, and the dynamics of age, race and ethnicity in much more detail than what could have been done 15 years ago. However, there are a lot of holes in the information. We do not know what causes the numbers to fluctuate. We know the context of the problems, but not the problem itself. Further analysis is needed desperately. The severe reduction of data analysts in the USFA has seriously hampered this effort. The information currently available is weak, and basic decisions on fire protection cannot be made.

The following lists, in priority, the other problem areas considered by Task Force 6.

1. Lack of funds and personnel to establish community relations services which allow fire service personnel to conduct prevention activities (proactive versus reactive fire protection).

2. More effective use of available time by fire fighters for fire prevention.

3. Lack of fire safety education in public and private schools, and a minimum standard for each grade level.

4. No public understanding of the hazards of fire and its cost to society.

5. Lack of funds for studying the impact of fire (i.e., a cost/benefit analysis of prevention activities).

6. Disproportionate impact of budget reductions on non-suppression forces (fire prevention).

7. No recognition by the fire service and the public of the scope of the fire problem or the value of fire prevention.

8. Lack of support for the development of new technology for automatic sprinkler systems which are acceptable to homeowners.

9. Inadequate data at both the national and local levels to document the extent and impact of fire problem; information is needed to define the problem.

10. Lack of federal incentives to develop and implement fire safety programs (e.g., fire-safe cigarettes, interior finishes).

11. No national criteria for fire prevention activities, namely, investigation, education, inspection and enforcement.

12. Reluctance by private and public sector organizations to accept fire protection/prevention methods (e.g., code and standard adoption and enforcement, and product combustibility/toxicity).

13. Need to regulate fire behavior of interior furnishings in elderly, child care and health care facilities.

14. Better sharing of information on fire prevention programs, techniques and training.

15. Lack of public appreciation of the value of maintenance inspections; public is unwilling to be more responsible for fire safety.

16. Limits placed by labor contracts on fire department prerogatives to assign personnel to fire prevention activities.

17. Lack of public fire safety education acceptance by fire service, educators and public.

18. Lack of fire safety education standard for teacher accreditation.

19. Range of personnel (both civilian and uniform) needed for fire prevention programs.

20. Federal government does not see fire problem as a national priority.

21. Lack of more comprehensive and expanded home fire safety programs.

U.S. Representative Curt Weldon (R-PA)

"The fire service is the heart and soul of America...the fire service forms the backbone of our country, and it has since our country was founded more than 200 years ago."

IV. Task Force Recommendations

The task force developed comprehensive solution strategies for the two critical issues. These are discussed below in priority order. While the task force identified a host of additional solution strategies, it did not have time to develop them fully. These candidate solutions are listed in Attachment 1.

1) Lack of knowledge of the fire problem and the value of fire prevention. The task force directed this charge against many actors in the fire protection community, including the fire service, the general public, educators and elected officials. The task force identified more than 25 programs to address this problem. The five highest in priority are listed below.

a. Initiate a mass media program to educate and influence the public. In conjunction with the National Advertising Council, the fire service would present, on a nationwide basis, the problems of fire and its tragic effects on life.

b. Join and support the Congressional Fire Services Caucus to educate and lobby national legislators.

c. Require that fire prevention education standards become a part of career paths for fire service personnel. Fifty percent of training time should be devoted to fire prevention. Service requirements should be at least two years active time for entry-level chief officers and three years active time for department chiefs.

d. Develop vivid and practical information for special groups, including the hotel/motel industry, handicapped, elderly and others at high risk.

e. Promote the adoption and application of residential sprinklers nationally.

2) Lack of support for a standardized reporting system to document needs/ problems and program results. The task force identified only eight recommended solutions to this problem. They follow in priority order.

a. Re-evaluate and reorganize the current national data system; a national task force under the aegis of the U.S. Fire Administration should be formed to review the operations of the National Fire Incident Reporting System. Specifically, the task force should be charged with reviewing the number and type of organizations which should contribute data, methods for disseminating information and reports, data collection methods, purposes and uses of data, hardware and software requirements, and scope of data base (whether EMS and hazardous materials data should be included).

b. The task force should identify total funding requirements and possible funding sources. Funding sources might include federal, state and local government organizations, industry grants, federal incentives and private (for profit) endeavors.

c. The task force should not have an open-ended agenda, and it should keep to its assigned schedule.

d. Upon the release of the task force's report, the fire service should support and continue to expand the established standardized national reporting system.

e. The USFA should publicize the data reporting system and all of its products, which should be distributed to the fire community in a timely manner. Further, the USFA should educate the fire protection community in the use of the data for decisionmaking at the local level.

f. Suggested reporting agencies for the standardized data system include the fire service, hospitals, doctors, medical examiners and industry. The objective is to develop a reporting system that has a broad constituency which not only will contribute to the system, but will be able to benefit from it.

g. Suggested reporting agencies should be represented on the task force.

h. The data system should be an organic one; the organization should provide training to the reporting agencies, a data feedback system and a vehicle for readily sharing information.

Attachment 1
Other Possible Solutions

1. *Collect objective data on the value of any prevention program.*

2. *Continue and enhance such current fire safety programs as "EDITH,, "Learn Not To Burn," 'Smoke Drills" and so forth.*

3. *Upgrade the national data system with in-depth investigation and analysis; dissemination of, and access to, this information should be augmented as well.*

4. *Educate the fire service to the need for changing its role (proactive versus reactive).*

5. *Increase visibility in public places, other than during emergencies and inspections.*

6. *Change fire service attitude to reward fire prevention accomplishments; suppression efforts should be regarded as failures.*

7. *Develop information and materials to be posted in public areas; seek the assistance of the private sector to help pay for development and placement costs.*

8. *Recruit non-traditional agencies for assistance, e.g., youth development, parks and planning.*

9. *Develop a Speakers bureau to take the fire problem to special interest groups.*

15. *Encourage new technology for automatic fire protection systems; maintain continuous contact between these industries and the fire service.*

16. *Solicitpublic official endorsements and support.*

17. *Develop publications for teachers on fire safety in the school and home.*

18. *Adopt code requirements mandating maintenance of fire protection systems.*

19. *Encourage fire prevention successes, especially for fire service personnel (e.g., increased recognition, financial incentives).*

20. *Urge all media to remind citizens to clean, test and repair smoke detectors at the beginning and end of daylight savings time.*

Task Force 7
Preparing for the 21st Century

———————————— Task Force Members: ————————————

Chief Michael Conley, International Fire Service Training Association, Stillwater (OK) -

Chief M.H. "Jim" Estepp, Prince George's County (MD) Fire Dept.

Dr. John Granite, St. James City (FL)

Mr. David A. Lucht, Worcester Polytechnic Institute

Mr. Martin Mintz, National Association of Home Builders

Chief Vic Porter, Berkeley (CA) Fire Department

Dr. Jack Snell, Center For Fire Research, National Institute for Standards and Technology

Mr. Harvey Ryland, Tampa (FL), Facilitator

I. Introduction

This section presents the results of Task Force 7, "Preparing for the 21st Century." The task force's objectives were: (1) to review the issues and recommendations presented in Chapters 18, 19 and 20 of *America Burning;* (2) to forecast the characteristics of the fire protection environment in the 21st century and the requirements and constraints associated with that environment; and (3) to prepare a general description of fire protection in the 21st century and recommend alternate concepts, policies, programs, techniques and activities for use by fire departments and other fire safety organizations in planning and preparing for the future.

II. Background

There have been significant changes in the concept and scope of fire protection since *America Burning* was published in 1973, and it is almost certain that even greater changes will occur as we enter the 21st century.

Some of the key changes that have occurred since 1973 include reduced resources for fire departments, resulting in closed companies or lower staffing levels; greatly expanded use of smoke detectors and automatic detection and suppression systems; significant increases in fire service involvement in emergency medical services, hazardous materials incidents and emergency management activities; revised personnel policies and schedules as a result of the Fair Labor Standards Act (FLSA); increased drug abuse in the fire service; and the creation of a federal fire effort implemented through the United States Fire Administration, National Fire Academy, and Center for Fire Research of the National Institute of Standards and Technology.

A "new" issue discussed in the task force was the increasing relationship between environmental protection, or the control of hazardous waste, and fire protection. In 1973, few fire officials even considered that harming the environment could be a spillover effect from fighting a fire. But it is a real possibility and underscores the complexities of modern life. For example, there have been incidents where contaminated water runoff resulting from suppressing a chemical plant fire polluted local groundwater. Environmental control agencies have expressed concern that the traditional fire fighting tactic of venting a building to release heat and smoke also may release hazardous gases. In addition, halon gases used in automatic suppression systems may be harmful to the ozone layer.

America Burning proposed the development of the federal fire effort and recommended the creation of the United States Fire Administration, National Fire Academy and Center for Fire Research, with a proposed annual budget of $124,840,000 (1973 dollars). The commission stated that the projected costs "can serve as an indication of minimum operating program needs and as a starting point for discussion." However, the maximum budget ever received by the U.S. Fire Administration was approximately $24 million, and there have been

several years in which there was a possibility that the United States Fire Administration budget would be zero. Consequently, many of the fire programs envisioned by the commission were curtailed because of inadequate funding.

The commission recognized that, as important as federal research is for combatting the nation's fire problems, the responsibility is not solely the government's. It recommended that the building materials and furnishings industry sponsor research directed toward improving fire safety in the built environment. While we have learned a lot in the past decade, much more information is needed to understand and control the fire hazards and risks posed by construction materials and building contents. Such associations as the Chemical Manufacturers Association, The Society of the Plastics Industry, Carpet and Rug Institute, International Association of Sleep Products, American Furniture Manufacturers Association, and Concrete and Masonry Institute have initiated studies to understand the combustion and toxicity hazards associated with their products.

Forecasting future trends, the commission attempted to plan for fire protection needs and demands in the 21st century. Examples of projected aspects of life included increased average life span, possibly to 120 years; extensive use of high-density housing; expanded adult and continuing education programs as people work until 100 years of age; automatic medical systems built into the body; use of exotic information and communications equipment; and a continued reduction in the fire suppression activities of fire departments. Such changes will have a major impact on fire protection in general, and on the fire service in particular.

Of course, even within a span of 15 years, many of the commission's individual conclusions and recommendations for the 21st century are no longer applicable or now may have a different priority. As the environment changes, so do the challenges and demands placed on the fire service. Updating *America Burning* will require a constantly revised picture of the future. We will need new information to identify upcoming fire protection conditions, requirements, constraints and priorities. This planning is critical if the fire service is going to meet tomorrow's demands.

III. Critical Issues

Chapters 18-20 of *America Burning* included the following topics: public and private involvement; basic and applied fire research; special research in human behavior, fire dynamics, and smoke and toxic gases; development of automatic detection and suppression systems; dissemination of results of research and development projects; private research activities; improvement of burn treatment; establishment of the U.S. Fire Administration and National Fire Academy; and operation of a national fire data system.

Task force members identified many issues which they. felt were most important in preparing for the 21st century. After consolidating these and ranking them in priority, the task force identified the following key issues (in priority order).

1) Cultural orientation. The single most critical factor to be considered is that public attitudes, behavior and values contribute significantly to our high fire losses. Examples of views and attitudes that people have which contribute to our relatively poor fire safety record (especially compared with other industrialized nations) include:

"it can't happen to me"

"odds are that it won't happen to me"

the insurance company will take care of me"

"it is not a disgrace to have a fire"

"I can set fires for revenge"

"people come out better after a fire"

"I will help those unfortunate people who had a fire"

"they're just children playing, they didn't know any better"

The consensus of the task force was that a high level of fire safety will not be achieved until these views and attitudes are changed.

2) Political action infrastructure, An organized and coordinated capability that can, at the national level, identify overall problems, establish priorities and "make things happen" (i.e., pass legislation, obtain budget increases) does not exist for fire protection. Unlike many other social and economic issues that have coordinated national representation, lobbying and public relations programs, fire safety appears to be a poor stepchild and only receives extensive public attention following a major incident.

The national representation for fire safety issues is fragmented, at best. The Joint Council of National Fire Service Organizations, the only national group concerned with fire safety, is composed primarily of fire service organizations. For fire safety to become a higher public

ing of buildings has increased the hazard and makes it critically important to know how a material will react in fire. Frequently, a new material is used without an understanding of how its properties react when heated or burned, until a tragedy occurs. Then, an after-the-fact research effort is initiated. If this understanding is achieved in advance, these properties could be predicted, and unsafe materials could be kept out of the built environment.

priority, a more comprehensive and broader coalition of interests must be formed that includes the active participation of all actors in the fire protection community. Associations and nationally recognized leaders of such professions as city and county managers, code and building officials, and architects/engineers must be a part of the fire safety solution.

3) Development of new and improved fire protection technology. Basic knowledge is needed in the physics and chemistry of combustion, especially concerning the generation of products of combustion and the process of extinguishment. Research needs to be done on the next generation of affordable, smart smoke detection and fire suppression systems.

Most fire deaths are from smoke inhalation, not heat or burns. Therefore, it is necessary to understand the components of smoke (i.e., products of combustion), including how they are generated, disseminated, and lead to injury and death. The increasing use of synthetic materials in the construction and furnish-

The general technology is available to support the development of advanced automatic smoke detection and fire suppression systems. However, budget restrictions at the National Institute of Standards and Technology's Center for Fire Research have limited the agency's ability to pursue aggressively the development of these systems. The center's involvement is essential because, apparently, the market does not provide sufficient incentives to encourage private industry to take on this task.

4) Fire protection information. Continuous and complete data collection and analysis to identify fire protection problems and solutions have been, and still are, top priorities. Unfortunately, the data collection and analysis efforts being conducted within the fire community are in disarray and not achieving the intended objectives. For example, while the U.S. Fire Administration has made significant progress with the National Fire Incident Reporting System (NFIRS), not all cities and states are participating in that network. Further, although this fire incident data is important, the fire service needs other types of information. Case studies of successful management techniques and new technologies that may have fire protection applications would be useful for fire managers.

There is not sufficient analysis of the information that is collected currently. The U.S. Fire Administration's budget restrictions have impeded the analysis and reporting of the NFIRS and other data collected by the National Fire Data Center. In recent years, personnel and contracting funds devoted to analyzing NFIRS data have been cut severely.

A comprehensive data collection and analysis capability would help the fire service. The information could be used to develop better fire spread models, predict the occurrence and location of fires, and establish improved tactical procedures. Also, this capability would support the infrastructure requirements discussed above. For example, better data might help in the adoption of improved fire safety legislation.

5) Redefine the traditional public safety service delivery system. The roles and responsibilities of the fire service must be redefined to meet the changing local government environment and the needs of the future.

A fundamental change is occurring in the nature and scope of what long has been considered the fire service's main responsibility, fighting fires. Within the past decade, there has been a downward trend in the number of structural fires being handled by fire departments. The National Fire Protection Association

estimates that the annual number of fires has decreased from slightly more than 1 million in 1977 to 800,000 in 1986. Plus, when the other *America Burning* recommendations are implemented, this traditional role of the fire department will be reduced and redefined further, although never eliminated.

The loss of federal General Revenue Sharing funds and other local government budget restrictions have reduced the resources of many departments, and this has affected company numbers and staffing levels.

Thus, the dual forces of limited resources and evolving demands for services are expected to have a major impact on the fire service. Few task force members felt that enough departments are prepared for these challenges, or are even in the process of preparing for them.

There are three key areas of fire service operations in need of examination.

a. What are the roles and responsibilities of the fire department of the future? During the past 20 years, the fire service has assumed a major role in providing emergency medical services (EMS), to the extent that such incidents now constitute a majority of some departments' runs. Fire department responsibility for hazardous materials protection also has been increasing, and with the role specified in the 1986 Emergency Preparedness and Community Right-To-Know Act, otherwise known as Title III of the Superfund Amendments and Reauthorization Act (SARA), this involvement will continue and possibly expand further. Many fire departments have increased their responsibility for building and environmental code enforcement, even in a few cases, to the extent of providing overall building department services.

b. What are the most effective and efficient methods for accomplishing these roles and responsibilities? It may be, for example, that a combination inspection/EMS/rescue/fire suppression unit that uses smaller apparatus will replace some of the standard engine companies in a community. Consolidated building and fire departments have proved to be an effective choice for many municipalities.

c. Is even the name "fire department" becoming obsolete? A few departments already are changing their names to reflect their expanding missions. For example, such names as the "Fire and Life Safety Department" and the "Department of Fire and Emergency Management" are becoming more common.

V. Task Force Recommendations

Upon reflecting on the impact of their deliberations, the members of the task force decided to issue only a limited number of recommendations. While all of the alternative solutions had merit and should be pursued, it was felt that the task force report would have greater impact if it were concise and targeted. Consequently, the task force's report consists of the five most important actions that need to be taken to improve fire protection in the future.

These recommended actions, in priority order, are:

1) Mount an on-going national campaign to change cultural orientation to one of fire safety consciousness, involving complete saturation of fire safety concepts to all age groups, using proven behavioral modification, marketing and simulation techniques.

2) Build and maintain a national fire safety constituency through the Congressional Fire Services Caucus, a fire safety political action committee (PAC), and a "conference board," composed of strong leadership from each public interest group and fire association, that will lobby on Specific issues.

3) Fund and begin the development of a new generation of affordable, automatic smoke detection and fire suppression systems, and initiate actions within the model code groups and legislatures around the county to install these systems in all dwelling units.

4) Develop and implement a widely accessible fire safety data base network, which will include the models and data needed for fire risk and hazard prediction and fire safety program productivity measurement. Hardware and software needed to access this data base should be provided to local fire departments.

5) Fund and conduct a comprehensive project to develop recommended future fire department roles and priorities and a new departmental name, with leadership provided by the Joint Council of National Fire Service Organizations in collaboration with elected and appointed local government officials and recognized national interest groups.

Attachment 1
Other Possible Solutions

The task force identified a number of alternative solution strategies for each of the critical issues just discussed. These alternatives are listed within each issue. The task force members did not assign a priority to these suggested solutions.

1) Cultural orientations

a. Fire safety public education programs should be mandated in schools for grades K through 12

b. Harsher legal penalties should be established for people whose carelessness results in a fire.

c. Negative cultural characteristics, which lead to unsafe fire practices, should be identified, and positive, safer behavior should be encouraged and reinforced.

d. A Mothers Against Drunk Driving (MADD) type program should be developed for citizen awareness and sensitivity toward negligent fires and those who cause them. This program should provide for positive intervention at every opportunity to change behavior, especially for "latch-key" children,

e. A symbol or theme for fire safety (e.g., Sparky) should be established which has national recognition and visibility similar to that achieved by "Spuds McKenzie."

f. Fire safety programs are often amateurish and less than effective. Experts skilled in advertising, promotion, and public relations need to be involved.

g. Government and other institutions can encourage fire safety by offering financial incentives (i.e., tax rebates or reductions) to those who do not have fires, practice fire safety behavior or install automatic fire protection systems.

h. The private sector needs to be an active participant in fire safety, which is good for business and society. The fire service needs to identify and use corporate sponsors for fire safety campaigns. (Note: The negative aspects of fire may preclude some companies from participating in such programs.)

i. Prominent national persons should be used in promotional materials and campaigns. Dick Van Dyke was very effective in the "Learn Not To Burn" spots .Football players could demonstrate "stop, drop and roll" for information targeted to younger children.

j. Neighborhood fire brigades should be developed to increase citizen awareness and education and provide first aid services.

k. A massive public education training program for fire service personnel, with support materials, should be developed and offered nationwide.

l. Media coverage should be obtained for fire department accomplishments, activities and such fire service events as the fire fighter Olympics.

m. Research should be conducted to determine what it takes to motivate members of each group in our society to improve its fire safety practices; special attention should be given to identifying positive techniques.

n. Quality fire protection television spot announcements should be developed to be shown on local stations and cable systems.

o. Model public education programs should be developed for specific ethnic and economic target groups.

p. Traffic tickets should be issued for fire protection code violations in structures other than single-family dwellings; fire fighters should have the authority to write tickets. Mandatory public service should be required to work off violations.

q. Provide graduate education (e.g., engineering and architecture) for fire safety professionals in all applicable fire protection disciplines.

2) Political action infrastructure

a. Everyone should ask their federal representatives and senators to join the Congressional Fire Services Caucus.

b. Establish a consolidated and comprehensive political action committee (PAC) to pursue fire safety programs.

c. Establish a fire protection "conference board" to lobby for specific issues. The "conference board" would be composed of strong leaders from each public interest group and fire protection professional association.

d. Establish a mechanism to educate local, state and national policymakers on fire protection requirements and issues.

e. Encourage the public to lobby for improved fire protection at local, state and national levels.

f. Provide incentives for fire protection organizations to work together, and disincentives for not working together.

g. The resources and influence of all local and national organizations (e.g., American Association of Retired Persons) should be used to improve fire protection.

3) Development of new and improved fire protection technology

a. Conduct needed fire research which is complete and thorough enough so the results can be put to end-item use.

b. Establish an effective mechanism for evaluating new fire protection engineering methods and disseminate results to potential users.

c. Institutionalize technology transfer through federal legislation requiring that fire service spinoffs be explored following major technological innovations in the behavioral, managerial or engineering sciences.

d. Continue and expand research into the combustion, toxicity and fire suppression phenomena.

e. Develop inexpensive automatic smoke detection and fire suppression systems that can: (1) recognize different types of fires; (2) direct suppressants toward the fire location; and (3) turn off when the fire is out.

4) Fire protection information

a. Establish a national laboratory network for fire cause analysis and investigation. Ensure that the information is readily accessible to local fire departments.

b. Establish a national fire information database network that, ideally, would be a combination of NFIRS, UFIRS, and non-incident information and ideas.

5) Redefine the traditional public safety service delivery system

a. Establish certification and licensing requirements for fire protection professionals.

b. Consolidate building and fire inspection services under fire departments.

c. Increase local fire department authority and responsibility to establish and accomplish goals.

d. Establish a significant national fire leadership training institute, e.g., expand the Harvard program to permit greater participation.

Appendix A. Project Steering Committee

Dr. Jack Snell
Center for Fire Research
National Institute of Standards and
   Technology
Gaithersburg, MD 20899
(301) 921-3143

Garry Briese
International Association of Fire Chiefs
1329 18th Street, N.W.
Washington, D.C. 20036
(202) 833-3420

Joseph O'Hagan
U.S. Army Corps of Engineers
Building 203, USAEA CA
Fort Myer
Arlington, VA 22211-5050
(202) 696-38 13

AI Whitehead
International Association of Firefighters
1750 New York Avenue, N.W.
Washington, D.C. 20006
(202) 737-8484

Tom Owens
Volunteer Coordinator
Fairfax County Fire and Rescue Department
4031 University Drive
Fairfax, VA 22032
(703) 691-2126

Harry Hickey
Johns Hopkins University
Applied Physics Laboratory
Johns Hopkins Road - Building 46
Laurel, MD 20707
(301) 792-5154

Howard Tipton
City Manager
P.O. Box 551
Daytona Beach, FL 32015(904) 252-6461

Gerard Hoetmer
International City Management Association
777 North Capitol Street, N.E.
Washington, D.C. 20002
(202) 962-4600

Deputy Chief Ray Alfred
District of Columbia Fire Department
1923 Vermont Avenue, N.W.
Washington, DC. 20001
(202) 462-1762

Philip Schaenman
Tri Data Corporation
1500 Wilson Boulevard
Arlington, VA 22209
(703) 841-2975

Rich Adams
Editorial Director
WUSA-TV
4001 Brandywine Street, N.W.
Washington, D.C. 20016
(202) 364-3905

Edward M. Wall
Deputy Administrator
U.S. Fire Administration
Emmitsburg, MD 21727
(301) 447-6771

James F. Coyle
Assistant Administrator
U.S. Fire Administration
Emmitsburg, MD 21727
(301) 447-6771

Alternates

Howard Markman
P.O. Box 177
Northfield, NJ 08225
(609) 645-7744

Colin A. Campbell
Colin A. Campbell & Associates
15-B East Windsor AvenueAlexandria, VA 22301
(703) 684-13 18

Hal Bruno
Political Director
ABC News
1717 DeSales Street, N.W.
Washington, D.C. 20036
(202) 887-7777

Appendix B.
America Burning Revisited Participants

Task Force

Louis J. Amabili 5
Director
Delaware State Fire School
RD 2, Box 166
Dover, DE 19901
(302) 736-4773

Deputy Chief Steven C. Bailey 1
President
National Fire Information Council
Seattle Fire Department
301 Second Avenue South
Seattle, WA 98104
(206) 386-1470

George B. Barney 4
Divisional Director
Engineering Services Division
Portland Cement Association
5420 Old Orchard Road
Skokie, IL 60077-4321
(312) 966-6200

Julius W. Becton, Jr.
Director
Federal Emergency Management Agency
Federal Center Plaza
500 C Street, S.W.
Washington, D.C. 20472
(202) 646-3923

Bob Blair
Chief, Intergovernmental Affairs
Federal Emergency Management Agency
Federal Center Plaza
500 C Street, S.W.
Washington, D.C. 20472
(202) 646-4600

Professor Richard E. Bland 1
435 East Irvin Avenue
State College, PA 16801
(814) 237-7262

Chief Paul Boecker 3
Chairman, Executive Board
International Fire Service Training
    Association
Lisle-Woodridge Fire District
1005 School Street
Lisle, IL 60532
(312) 964-2233

Clyde A. Bragdon, Jr.
Administrator
U.S. Fire Administration
16825 South Seton Avenue
Emmitsburg, MD 21727
(301) 447-1080

Tom Brennan 2
Editor
Fire Engineering Magazine
250 Fifth Avenue
New York, NY 10001
(212) 481-5771

Garry L. Briese 3
Executive Director
International Association of Fire Chiefs
1329 18th Street, N.W.
Washington, D.C. 20036
(202) 833-3420

Peter Brigham 1
President
Burn Foundation
1311 Chancellor Street
Philadelphia, PA 19107
(215) 735-4060

David Goldston
Technical Consultant
Science, Research and Technology
    Subcommittee
U.S. House of Representatives
2319 Rayburn House Office Building
Washington, D.C. 20515
(202) 225-8844

Dr. John Granito 7
2961 Bowsprit Lane
St. James City, FL 33956
(813) 283-2438

Olin Greene 5
State Fire Marshal
3000 Market Street Plaza
Suite 534
Salem, OR 97310-0198
(503)378-FIRE

William H. Hansell, Jr. 3
Executive Director
International City Management Association
777 North Capitol Street, N.E.
Washington, D.C. 20002
(202) 962-4600

John Glenn Hart, III 1
Assistant Administrator
U.S. Fire Administration
16825 South Seton Avenue
Emmitsburg, MD 21727
(301) 447-1105

Charles A. Henry
Fire Commissioner
Commonwealth of Pennsylvania
Box 3321
Harrisburg, PA 17105-3321
(717) 783-5120

Tom Hem 3
International Association of Fire Fighters
1750 New York Avenue, N.W.
Washington, D.C. 20006
(202) 737-8484

Gerard Hoetmer 5
Staff Associate
International City Management Association
777 North Capitol Street, N.E.
Washington, D.C. 20002
(202) 962-4600

Chief Warren E. Isman   2
President
International Association of Fire Chiefs
Fairfax County Fire and Rescue Department
4031 University Drive
Fairfax, VA 22030
(703) 691-2546

Rolf Jensen, P.E. 4
President
Society of Fire Protection Engineers
60 Batterymarch Street
Boston, MA 02110
(312) 948-0700

Matt Kane 1
Staff Associate
National League of Cities
1301 Pennsylvania Avenue, N.W.
Washington, DC. 20001
(202) 626-3000

Assistant Chief Charles H. Rime 6
Fire 'Marshal
Phoenix Fire Department
520 West Van Buren Street
Phoenix, AZ 85003
(602) 262-6002

Ms Janet Kimmerly 4
Editor
Firehouse Magazine
210 Crossways Park Drive
Woodbury, NY 11797
(516) 496-8000

Chief Jerry Knight 3
St. Petersburg Fire Department
1429 Arlington Avenue, North
St. Petersburg, FL 33705
(813) 893-7694

Richard P. Kuchnicki 4
President
Council of American Building Officials
5203 Leesburg Pike
Falls Church, VA 22041
(703) 931-4533

Chief Richard Lamb 5
National Volunteer Fire Council
Wayne Township Fire Department
6456 West Ohio Street
Indianapolis, IN 46224
(317) 247-8501

Professor David A. Lucht 7
Director
Center for Firesafety Studies
Worcester Polytechnic Institute
Worcester, MA 01609
(617) 793-5593

Ms Barbara Lundquist 1
Tri Data Corporation
1500 Wilson Boulevard
Arlington, VA 22209
(703) 841-2875

Arnold Luterman, M.D 6
Director, Burn Center
Ripps-Meisler Professor of Surgery
University of South Alabama
2451 Fillingim Street
Mobile, AL 36617
(205) 47 1-7049

Hugh McClees
Aspen Systems Corporation
1600 Research Boulevard
Rockville, MD 20850
(301) 251-5159

Edward H. McCormack 6
Executive Director
International Society of Fire Service
    Instructors
20 Main Street
Ashland, MA 01721
(617) 881-5800

Chief Roger McGary 2
President
International Society of Fire Service
    Instructors
Takoma Park Fire Department
7201 Carroll Avenue
Takoma Park, MD 21215 ....
(301) 270-4242

John McNichol
Legislative Assistant
U.S. Representative Curt Weldon
1233 Longworth House Office Building
Washington, DC. 20515
(202) 225-5222

John Marshall
Fire Program Specialist U.S. Fire
    Administration
16825 South Seton Avenue
Emmitsburg, MD 21727
(301) 447-1122

Martin Mintz 7
Senior Technical Advisor
National Association of Home Builders
15th and M Streets, N.W.
Washington, D.C. 20005
(202) 822-0200

Chief Richard Moreno 6
Tucson Fire Department
P.O. Box 272210
Tucson, AZ 85726-7210
(602)791-4511

Patrick Murphy 6
U.S. Conference of Mayors
1620 I Street, N.W.
Washington, DC. 20006
(202) 293-7330

William M. Neville, Jr. 2
Superintendent
National Fire Academy
16825 South Seton Avenue
Emmitsburg, MD 21727
(301) 447-1123

J. Thomas Smith
Fire Fighter Health and Safety Specialist
U.S. Fire Administration
Federal Center Plaza
500 C Street, S.W.
Washington, DC. 20472
(202) 646-2449

Fire Marshals Association of North America
Suite 1210
1110 Vermont Avenue, N.W.
Washington, D.C. 20005
(202) 667-7441

Director
Center for Fire Research
U.S. Department of Commerce
National Institute of Standards and
   Technology
Gaithersburg, MD 20899
(301) 975-6850

Vice President - Public Affairs
The Tobacco Institute
706 Upham Place
Vienna, VA 22180
(202) 457-9313

Hartford Fire Department
275 Pearl Street
Hartford, CT 06103
(203) 722-8200

Associate Director for Public Safety Programs
National Criminal Justice Association
444 North Capitol Street, N.W.
Suite 608
Washington, D.C. 20001
(202) 347-4900

U.S. Representative Doug Walgren
Chairman
Science, Research and Technology
   Subcommittee
2319 Rayburn House Office Building
Washington, D.C. 20515
(202) 225-8844

Edward M. Wall
Deputy Administrator
U.S. Fire Administration
16825 South Seton Avenue
Emmitsburg, MD 21727
(301) 447-1080

U.S. Representative Curt Weldon
Chairman
Congressional Fire Services Caucus
1233 Longworth House Office Building
Washington, D.C. 20515
(202) 225-5222

President
International Association of Black
   Professional Fire Fighters
P.O. Box 22005
Seattle, WA 98122
(206) 329-9870

Appendix C.
Status Report on the 90 Recommendations from America Burning

Section I
————Introduction————

One of the major events of the United States fire service occurred in June 1973 - the publication of *America Burning*. This document is the report of the National Commission on Fire Prevention and Control, a presidential commission appointed by President Richard M. Nixon. Richard E. Bland (associate professor, Pennsylvania State University) was commission chairman, and W. Howard McClennen (president of the International Association of Fire Fighters) was vice chairman. Commission members included members of Congress, the Secretaries of the Departments of Commerce and Housing and Urban development, and representatives of public and private organizations concerned with fire protection, including the fire service, insurance industry, news media, academia and the building industry.

The commission conducted research into the U.S. fire problem, held a series of hearings throughout the country and deliberated the solutions to these problems.

The result of this two-year effort was the commission's report, *America Burning*.

This report contains 90 recommendations concerning the improvement of fire protection in the U.S. These recommendations cover the following general areas:

* burn prevention and treatment;

* fire fighter health and safety;

* building and fire codes and standards;

* automatic detection and suppression;

* fire protection master planning;

* fire department organization and operation;

* rural and wildland fire protection;

* public education;

* fire prevention inspection and enforcement;

* incentives for improved fire safety;

* transportation fire safety; and

* establishment of federal organizations for fire protection research, data collection and analysis, planning and training.

In the 15 years since *America Burning* was published; many of the recommendations have been accomplished, and other recommendations have been accomplished partially. Some recommendations have been attempted without success, and others apparently have not been attempted at all.

The most visible accomplishment was the creation of the U.S. Fire administration, National Fire Academy and Center for Fire Research at the National Institute of Standards and Technology. However, some fire protection leaders have stated that this is only a partial success because the funding for these organizations has never come close to the amounts recommended by the commission.

Significant successes also have been achieved in such areas as installation of smoke detectors, improvement of fire and building codes, and fire protection master planning. Some success has been achieved for other recommendations, e.g., expanding public education programs and increasing fire department involvement in fire prevention. Other recommendations were not included in the legislation resulting from America Burning (P.L. 93-498) and, therefore, have not been implemented. For example, recommendations pertaining to fire department grants for training and equipment were not included in the legislation. Thus, such recommendations have not been accomplished at all.

The remainder of this report consists of the following sections:

Section II. Recommendation and Accomplishments - lists each of the 90 recommendations contained in America Burning along with a discussion of the extent to which each has been accomplished.

Section III. Summary of Accomplishments - contains a general discussion of (1) recommendations which have been substantially accomplished; (2) recommendations that have not been accomplished to any practical degree; and (3) recommendations for which significant progress has been made, but continued effort is still required.

Section II
Recommendations and Accomplishments

2.1 Chapter 1 The Nation's Fire Problem

Recommendation

1   The commission recommends that Congress establish a U.S. Fire Administration to provide a national focus for the nation's fire problem and to promote a comprehensive program with adequate funding to reduce life and property loss from fire.

Accomplishments

The United States Fire Administration (USFA) was established in 1975 as the National Fire Prevention and Control Administration. Also established were the National Fire Academy (NFA), originally a unit of the fire administration, and the Center for Fire Research (CFR), a unit of the National Institute of Standards and Technology. The commission recommended an average annual budget . for the first five years of $125 million. However, the largest annual budget ever received was less than $24 million, including funding for the CFR.

Recommendation

2.. The commission recommends that a national fire data system be established to provide a continuing review and analysis of the entire fire problem.

Accomplishments

The National Fire Data Center (NFDC) was established in 1975 as a major component of the USFA. The National Fire Incident Reporting System (NFIRS) was implemented as the foundation of the NFDC, and now has 37 states and 20 metropolitan areas reporting data. Currently, the NFDC is conducting general data analysis, as well as special studies of fire fighter and residential fire facilities, and major and unusual fires. In addition, the NFDC is conducting a project to improve fire department long-range planning and tactical decisionmaking capabilities through automated management information systems.

2.2 Chapter 2 Living Victims of the Tragedy

Recommendation

3.. The commission recommends that Congress enact legislation to make possible the attainment of 25 burn units and centers and 90 burn programs within the next 10 years.

Accomplishments

Not included in legislation (P.L. 93-498), thus, there are no corresponding programs. However, by 1983, there were approximately 125 burn centers in the United States, including one in virtually every metropolitan area with a sufficient population base, a significant achievement. Such federal actions as the enactment of Medicare and Medicaid, and

support of medical research and training have contributed to this progress.

Recommendation

4.. The commission recommends that Congress, in providing for new burn treatment facilities, make adequate provision for the training and continuing support of the specialists to staff these facilities. Provision also should be made for the special training of those who provide emergency care for burn victims in general hospitals.

Accomplishments

Not included in legislation (P.L. 93-498), thus, there are no corresponding programs. However, through other resources, this training is being carried out now on a broad scale by professional organizations and emergency medical service programs, and by burn centers throughout the nation.

Recommendation

5.. The commission recommends that the National Institutes of Health greatly augment their sponsorship of research on burns and burn treatment.

Accomplishments

In the mid-1970s, the National Institute of General Medical Sciences granted funds for burn research programs at seven burn center hospitals and other public and private grants supported research at other burn centers.

Recommendation

6.. The commission recommends that the National Institutes of Health administer and support a systematic program of research concerning smoke inhalation injuries.

Accomplishments

Not included in legislation (P.L. 93-498), thus, there are no corresponding programs. Also, see recommendation number 35.

2.3 Chapter 3 Are There Other Ways?

Recommendation

7.. The commission recommends that local governments make fire prevention at least equal to suppression in the planning of fire department priorities.

Accomplishments

In general, fire prevention efforts have increased dramatically, including expanded bureau and in-service company inspections and enforcement, improved public education programs, and increased involvement in land use planning and building/fire code development. The USFA, NFPA and other fire protection organizations have conducted numerous projects to promote and support the expansion of fire prevention activities by local departments. While these projects have helped to increase the amount of local government resources devoted to fire prevention, the goal of making "fire prevention at least equal to suppression" is yet to be reached, except in a few departments. For example, the Visalia (California) Fire Department currently devotes approximately 60% of its annual budget to prevention.

Recommendation

8.. The commission recommends that communities train and use women for fire service duties.

Accomplishments

The number of women in the fire service has increased significantly over-the last 14 years. The first career woman fire fighter was hired by the Arlington County (Virginia) Fire Department in 1974 and is now a captain with that department. The USFA promoted the inclusion of women in the fire service and provided information to assist fire departments in recruiting and retaining women for fire fighting duties. This USFA program included the conduct of a conference on women in the fire service (1981), and preparation of the reports, Role of Women in the Fire Service, and Issues for Women in the Fire Service. The USFA is currently developing protective clothing and equipment sizing information for use by manufacturers in supplying items especially designed for female fire fighters.

Recommendation

9.. The commission recommends that laws which hamper cooperative arrangements among local fire jurisdictions be changed to remove the restrictions.

Accomplishments

There are no known statistics on the number of mutual (including automatic) aid agreements that have been implemented over the past 14 years, or revision of associated laws. However, the use of such agreements has been promoted by the USFA (through its Integrated Emergency Management System [IEMS] project, for example), and by conducting a fire service seminar on mutual aid. It is felt that the number and extent of use of mutual aid agreements have increased. The budget limitations experienced by local governments probably have contributed to this increase.

Recommendation

10.. The commission recommends that every local fire jurisdiction prepare a master plan designed to meet the community's present and future needs in fire protection, to serve as a basis for program budgeting, and to identify and implement the optimum cost-benefit solutions in fire protection.

Accomplishments

Numerous fire departments have prepared fire protection master plans as a result of the commission's recommendation and a comprehensive program conducted by the USFA. This program included the development of the master planning process (for both single and multi-jurisdictional planning), preparation of a set of manuals for use by departments in preparing plans, a national conference on master planning, a report to Congress on master planning, training courses and a planning support team. The number of departments that have prepared a master plan is unknown, but it is known that many plans have been prepared and some departments have updated their plans several times.

Recommendation

11.. The Commission recommends that federal grants for equipment and training be available only to those fire jurisdictions that operate from a federally approved master plan for fire protection.

Accomplishments

The legislation (PL 93-498) did not provide for grants for training and equipment.

Recommendation

12.. The commission recommends that the proposed U.S. Fire Administration act as a coordinator of studies of fire protection methods and assist local jurisdictions in adapting findings to their fire protection planning.

Accomplishments

The USFA has served as a clearinghouse for fire protection studies, ideas and information, and has disseminated such information through the Learning Resource Center and such publications as the arson and public education resource exchange bulletins.

2.4 Chapter 4 Planning for Fire Protection

Recommendation

13.. The commission recommends that the proposed U.S. Fire Administration provide grants to local fire jurisdictions for developing master plans for fire protection. Further, the proposed U.S. Fire Administration should provide technical advice and qualified personnel to local fire jurisdictions to help them develop master plans.

Accomplishments

The legislation (PL 93-498) did not provide for grants to fire departments for master planning. The USFA provided on-site master planning technical assistance during the period 1978-1982. This support was discontinued when the USFA was reorganized in 1983.

2.5 Chapter 5 Fire Service Personnel

Recommendation

14.. The commission recommends that the proposed U.S. Fire Administration sponsor research in the following areas: productivity measures of fire departments, job analyses, fire injuries and fire prevention efforts.

Accomplishments

Numerous projects were conducted within these areas, especially fire fighter safety and fire prevention. The major productivity projects included a fire department working relationships study and the analysis of alternative company staffing levels. The apparatus staffing study was not completed because of USFA budget limitations. Currently, the USFA is conducting a study of alternative methods of providing fire protection, with results being prepared for dissemination to state and local governments.

Recommendation

15.. The commission urges the federal research agencies, for example, the National Science Foundation and the National Institute of Standards and Technology, to sponsor research appropriate to their respective missions within the areas of productivity of fire departments, causes of fire fighter injuries, effectiveness of fire prevention efforts and the skills required to perform various fire department functions.

Accomplishments

The primary effort in these areas was accomplished by the USFA (see #14 above) because the missions of these agencies evolved into more technical issues. The USFA assumed responsibility for some of the fire projects initiated by NSF and NIST, for example, the pumping apparatus specifications and fire protection master planning projects. Projects conducted by the USFA in these 'areas resulted in the preparation of the following documents (as examples): Model Performance Criteria for Structural Fire Fighters' Helmets, Development of a Job-Related Physical Performance Examination for Fire Fighters - A Summary Report, Survey of Fire Fighter Injuries (in cooperation with IAFF), and National Fire Service System./ Task Analysis Phase I: Development of the Analysis Process. The USFA currently is sponsoring a fire fighter mortality study being conducted by the University of Washington.

Recommendation

16.. The commission recommends that the nation's fire departments recognize advanced and specialized education, and hire or promote persons with experience at levels commensurate with their skills.

Accomplishments

Many fire departments have recognized that specialized education and advanced degrees can be of benefit to the department, and have included such criteria in recruiting or promoting for specific positions.

Recommendation

17.. The commission recommends a program of federal financial assistance to local fire services to upgrade their training.

Accomplishments

The NFA currently provides financial assistance to local fire services through both resident and field training programs.

This assistance includes student stipends for resident classes, "train the trainer" (provides for training of state and local training personnel in the delivery of courses prepared for hand-off to state or local training organizations), direct supplementary delivery (supplements state and local training efforts), Academy Planning and Assistance Program (provides technical and training assistance to state and local fire service organizations), and Open Learning for the Fire Service (provides fire service personnel with the opportunity to earn baccalaureate degrees in fire administration or fire technology).

Recommendation

18.. In the administering of federal funds for training or other assistance to local fire departments, the commission recommends that eligibility be limited to those departments that have adopted an effective, affirmative action program related to the employment and promotion of members of minority groups.

Accomplishments

This recommendation has been accomplished through the Federal Procurement Regulations because these regulations have equal opportunity requirements applicable to grantees.

Recommendation

19.. The commission recommends that fire departments, lacking emergency ambulance, paramedical and rescue services, consider providing them, especially if they are located in communities where these services are not provided adequately by other agencies.

Accomplishments

Numerous fire departments have implemented and expanded EMS pro-

grams over the past 14 years. The USFA has conducted a number of activities to promote the establishment of fire service EMS programs, for example, the "National Workshop for Fire Service EMS Needs - the Rockville Report." In addition, the EMS standards established by DOT for personnel, vehicles and equipment have had a major impact on improving EMS programs at the local level. However, there are communities which still do not have adequate EMS programs.

2.6 Chapter 6 A National Fire Academy

Recommendation

20.. The commission recommends the establishment of a National Fire Academy to provide specialized training in areas important to the fire services and to assist state and local jurisdictions in their training programs. Accomplishments

The National Fire Academy has been operational since 1975 and has operated the fire training facility at Emmitsburg since January 1980. The academy has numerous programs to provide specialized training (hazardous materials, for example) and has provided assistance to local training activities, including course development and "train the trainer" courses.

Recommendation

21.. The commission recommends that the proposed National Fire Academy assume the role of developing, gathering and disseminating, to state and local arson investigators, information on arson incidents and on advanced methods of arson investigations.

Accomplishments

The NFA conducts arson investigation and related training. The USFA maintains an arson research and information dissemination effort, with the fire data center collecting and analyzing information on arson incidents. Currently, the USFA is developing a community-based organization anti-arson program, a juvenile firesetter program and an Arson Information Management System, and is disseminating the results of a study of the needs of the rural arson investigator. The previously published Arson Resource Directory is being updated, and the Arson Resource Center, located at the Learning Resource Center, is now computerized for rapid access and updating. The CFR compiled the information, edited and published a Fire Investigation Handbook that is in widespread use among arson investigators.

Recommendation

22.. The commission recommends that the National Fire Academy be organized as a division of the proposed U.S. Fire Administration which would assume responsibility for deciding details of the academy's structure and administration.

Accomplishments

During the period 1975-1981, the NFA was a component of the USFA. In 1981, the NFA was separated administratively from the USFA.

Recommendation

23.. The commission recommends that the full cost of operating the proposed National Fire Academy and subsidizing the attendance of fire service members be borne by the federal government.

Accomplishments

The NFA pays up to 75% of the cost of attending courses at the academy. Generally, the only cost to the attendee is a nominal payment for mealsand for local transportation at the departure location. This financial subsidy even includes airfare for attendees. However, continued subsidy is dependent on the availability of corresponding funds in the NFA budget.

2.7 Chapter 7 Equipping the Fire Fighter

Recommendation

24.. The commission urges the National Science Foundation, in its Experimental Research and Development Incentives Program, and the National Institute of Standards and Technology, in its Experimental Technology Incentives Program, to give high priorityto the needs of the fire services.

Accomplishments

The National Science Foundation stopped its applied fire research activities upon establishment of the USFA and CFR. The NIST Experimental Technology Incentives Program did conduct a major project on methods for fire retarding polyester/cotton apparel fabrics before the ETIP program was stopped. The CFR program has included several projects in response to the needs of the fire services, for example, fire fighter turnout coats, helmets, lightweight air tanks and methodology for locating fire stations.

Recommendation

25.. The commission recommends that the proposed U.S. Fire Administration review current practices in terminology, symbols and equipment descriptions, and seek to introduce standardization where it is lacking.

Accomplishments

The USFA, NFA and CFR have promoted several areas of standardization, including the National Fire Incident Reporting System, fire fighter protective clothing and equipment, the Incident Command System, terminology and through training courses at the NFA.

Recommendation

26.. The commission urges rapid implementation of a program to improve breathing apparatus systems and expansion of the program's scope where appropriate.

Accomplishments

Tremendous progress has been made in the design of safer, more effective breathing apparatus. In the early 1970s the NIST conducted a research and development project on the design of higher pressure, lighter weight air tanks. The USFA's Project FIRES, conducted in cooperation with the National Aeronautics and Space Administration (NASA), produced a new generation of breathing apparatus which is now widely available and used. The NFPA, under a USFA grant, prepared the Manual for Selection, Use, Care and Maintenance of Fire Fighter Self-Contained Breathing Apparatus. In addition, a USFA project with the Bureau of Mines addressed the design and testing of a (closed circuit) "Low Profile Rescue Breathing Apparatus." A prototype long-duration (two-hour), positive-pressure oxygen-breathing apparatus has been developed and is being field tested

Recommendation

27.. The commission recommends that the proposed U.S. Fire Administration undertake a continuing study of fire service equipment needs, monitor research and development in progress, encourage needed research and development, disseminate results and provide grants to fire departments for equipment procurement to stimulate innovation in equipment design.

Accomplishments

To date, the USFA's equipment research and development activities have been focused on fire fighter protective equipment, smoke detectors and reporting systems, automatic detection and suppression systems, and specifying pumping apparatus. All of these activities have resulted in the improvement of equipment design and operation, as well as the dissemination of corresponding information. The USFA and NASA have developed and are testing a "hands-free" communications device which is fully compatible with self-contained breathing apparatus. The USFA legislation did not provide for grants for equipment procurement. A new portable monitor that will quickly detect and identify hazardous chemical vapors is being field tested by fire departments under USFA sponsorship. The USFA is cooperating with the U.S. Coast Guard in the lab testing of chemical protective clothing, and with the Coast Guard and the Department of Energy in evaluating a hazardous chemical protective ensemble.

Recommendation

28.. The commission urges the Joint Council of National Fire Service Organizations to sponsor a study to identify shortcomings of fire fighting equipment and the kinds of research, development or technology transfer that can overcome the deficiencies,

Accomplishments

Members of the joint council have contributed to numerous USFA and NFA activities, especially as members of project advisory committees and the NFA Board of Visitors. However, the specified study was not conducted. The USFA did conduct a project to establish fire protection research and development priorities.

2.8 Chapter 8 No Recommendations

2.9 Chapter 9 The Hazards Created Through Materials

Recommendation

29.. The commission recommends that research in the basic processes of ignition and combustion be strongly increased to provide a foundation for developing improved test methods.

Accomplishments

The CFR conducts, both in-house and through a research grants program, fundamental and applied research in the physics and chemistry of fire leading to improved methods for measuring fire properties and fire performance of materials and products. The work ranges from flame chemistry and polymer degradation to the development of measurement and predictive methods. The fundamental research is used to underpin the more applied work. The research in flame chemistry is leading to the ability to predict the evolution of gases and particu-

lates from the flame, e.g., soot and carbon monoxide. The work in polymer degradation is leading to the ability to design and produce more fire-safe materials. Measurement methods are being developed that will provide the scientifically based data needed as input to the predictive models.

Recommendation

30.. The commission recommends that the new Consumer Product Safety Commission (CPSC) give a high priority to the combustion hazards of materials in their end use.

Accomplishments

The CPSC, with technical backup from the CFR, has addressed the fire hazard of many consumer products by establishing mandatory standards, promoting voluntary standards, urging product recalls and providing consumer information. Included products are children's sleepwear, general apparel, mattresses, upholstered furniture, blankets, fire-safe cigarettes, cigarette lighters, matches, fireworks and heating equipment.

Recommendation

31.. The commission recommends that the present fuel load study sponsored by the General Services Administration and conducted by the National Institute of Standards and Technology be expanded to update the technical study of occupancy fire loads.

Accomplishments

The CPR conducted the fuel loads study for the GSA. The development by CFR of methods to predict the growth and spread of fire from measured fuel characteristics indicates the need for new surveys to provide information for fire hazards and risk calculations.

Recommendation

32.. The commission recommends that flammability standards for fabrics be given high priority by the Consumer Product Safety Commission.

Accomplishments

The NIST and CPSC have implemented four flammable fabric standards, two which are mandatory and two which are voluntary in cooperation with manufacturers and trade associations. The children's sleepwear standard is mandatory. Prior to its adoption, there were approximately 60 child sleepwear-related fire deaths per year; now this number is approximately two deaths per year. A voluntary nightwear standard is being prepared in cooperation with industry. This standard will provide for point of sale comparative information on the flammability of various fibers. At the present time, approximately 75% of apparel fire deaths are persons over 66 years of age. This consumer information program should help to reduce that number. The Mattress Flammability Standard is mandatory and, as a result, almost every mattress being produced today will resist ignition by cigarette. The value of this standard is demonstrated by the fact that, during the period 1980 - 1984, there was an approximately 32% reduction in cigarette-ignited mattress fire deaths. In 1977, industry accepted a voluntary standard for upholstered furniture. Before this standard, approximately 10-15% of upholstered furniture would resist cigarette ignition; now approximately 68% will resist such

ignition. During the period 1980 - 1984, cigarette-ignited upholstery fires decreased by approximately 24%.

Recommendation

33.. The commission recommends that all states adopt the Model State Fireworks Law of the National Fire Protection Association, thus prohibiting all fireworks except those for public displays.

Accomplishments

The NFPA, USFA and other organizations have promoted the adoption of this law, and many states have adopted this or similar laws.

Recommendation

34.. The commission recommends that the Department of Commerce be funded to provide grants for studies of the dynamics of combustion and the means of its control.

Accomplishments

The National Science Foundation RANN program on fire research was transferred to the CFR (a part of the National Institute of Standards and Technology which is an agency of the Department of Commerce) in the mid-1970s. CFR added funds from its base appropriation to increase the grants program to $2 million per year and annually funds more than 20 grants, mostly at universities. The CFR grants program is an integral part of the CFR program and is a way to bring the best scientific expertise in many disciplinary fields to focus on the fire problem. (See recommendation number 29 for a further description of the CFR program.)

Recommendation

35.. The commission recommends that the National Institute of Standards and Technology and the National Institutes of Health cooperatively devise and implement a set of research objectives designed to provide combustion standards for materials to protect human life.

Accomplishments

CFR has conducted a broad program of research into such fire problem areas as ignition, flame spread, products of combustion, extinguishment, detection and heat release. This program has led to the establishment or modification of many standards, mostly through the consensus standards-setting process, for example, the CPSC children's sleepwear, mattress and insulation standards; the flooring radiant panel test for floor coverings adopted by ASTM, NFPA and ISO; testing and installation standards for smoke detectors adopted by NFPA, UL and others; and the flame spread test for wall lining materials adopted by the International Maritime Organization. However, standards have not specifically been developed in cooperation with the National Institutes of Health.

2.10 Chapter 10 Hazards Through Design

Recommendation

36.. The commission urges the National Institute of Standards and Technology to assess current progress in fire research and define the areas in need of additional investigation. Further, the institute should recommend a program for translating research results into a systematic body of engineering principles and, ultimately, into guidelines useful to code writers and building designers.

126   Appendix C

Accomplishments

The CFR prepares both short- and long-term research plans based on an assessment of current and future needs of fire research results. In addition, CFR has established the National Fire Research Strategy Conference to prepare and continually update a fire research strategy for the nation. A major CFR thrust has been the translation of research results into a systematic body of engineering knowledge to provide guidelines, formulae, models, etc., to establish a soundly based fire protection engineering profession. Progress includes a set of engineering formulae known as FIRE-FORM, a prototype fire hazard assessment method called Hazard I, and several engineering models of varying sophistication that calculate fire growth and smoke spread.

Recommendation

37.. The commission recommends that the National Institute of Standards and Technology, in cooperation with the National Fire Protection Association and other appropriate organizations, support research to develop guidelines for a systems approach to fire safety in all types of buildings.

Accomplishments

The systems approach to building fire safety continues as a major project in the CFR. An important series of developments were the Fire Safety Evaluation Systems for various occupancies adopted in consensus standards and/or in various regulations. A more scientifically based system is currently in development.

Recommendation

38.. The commission recommends that, in all construction involving federal money, awarding of those funds be contingent upon the approval of a fire safety systems analysis and a fire safety effectiveness statement.

Accomplishments

A "Study of Fire Safety Effectiveness Statements" was conducted by the USFA as part of this effort, with the conclusion that liability issues would preclude the practical application of this concept. The General Services Administration now uses the fire safety systems analysis concept in designing new federal buildings.

Recommendation

39.. The commission urges the Consumer Product Safety Commission to give high priority to matches, cigarettes, heating appliances and other consumer products that are significant sources of burn injuries, particularly products for which industry standards fail to give adequate protection.

Accomplishments

The CPSC, with technical support from the CFR, has addressed fire safety issues associated with matches, cigarettes, cigarette lighters and heating equipment, resulting in the adoption and use of mandatory requirements for matchbooks; the completion of a study of the technical and commercial feasibility of producing a cigarette that would not ignite material; the initiation of a cigarette lighter project which includes a study of how lighters start fires (approximately 200 people die each year from fires caused by lighters) and the consideration of lighter designs which are more difficult for use by children (i.e., "child proof"); and a study of electric, gas, kero-

sene and wood heaters to identify ways to improve design and maintenance. The CPSC has issued a regulation requiring labels on wood stoves which provides critical consumer information on installation, use and maintenance. (Approximately 20% of residential fires are caused by wood heating equipment.)

Recommendation

40.. The commission recommends to schools giving degrees in architecture and engineering that they include in their curricula at least one course in fire safety. Further, we urge the American Institute of Architects, professional engineering societies and state registration boards to implement this recommendation.

Accomplishments

The CFR has been instrumental in establishing courses in fire science and fire safety engineering at several small colleges and universities. The referenced organizations have not implemented the recommendation, however.

Recommendation

41.. The commission urges the Society of Fire Protection Engineers to draft model courses for architects and engineers in the field of fire protection engineering.

Accomplishments

The accomplishment in this area was achieved through a USFA grant to the Society of Fire Protection Engineers to prepare a methodology for fire-safe building design. The results of this project were documented in the report, Document the Final Fire Safety Methodology. Referenced model courses have yet to be developed.

Recommendation

42.. The commission recommends that the proposed National Fire Academy develop short courses to educate practicing designers in the basis of fire safety design.

Accomplishments

The NFA has developed and offers resident, hand-off and Open Learning courses in fire-safe building design.

2.11 Chapter 11 Codes and Standards

Recommendation

43.. The commission recommends that all local governmental units in the United States have in force an adequate building code and fire prevention code or adopt whichever they lack.

Accomplishments The USFA and NFA have promoted the adoption of building/ fire codes and the strengthening of existing codes. While quantitative data is not available, it is known that many communities have adopted new codes or improved existing codes. For example, many communities have adopted automatic detection and suppression system, smoke detector, fireworks and/or noncombustible roof ordinances. The NFA offers resident and field courses that address code development. There are still many communities that do not have adequate codes, and some do not have codes of any kind.

Recommendation

44.. The commission recommends that local governments provide the competent personnel, training programs for inspectors and coordination among the various departments involved to enforce effec-

tively the local building and fire prevention codes. Representatives from the fire department should participate in reviewing the fire safety aspects of plans for new building construction and alterations to old buildings.

Accomplishments

Building/fire code enforcement has improved significantly as fire departments have expanded their fire prevention programs (see Recommendation #7). Local and state governments (Florida, for example) have established training requirements for company and fire prevention bureau personnel who will be involved in inspection/enforcement activities. Relationships between fire and building departments have been enhanced in many communities, and, in some cases, the fire department has an inspector assigned to the building department. Furthermore, fire departments have become more involved in land use planning and review of proposed developments. The master planning program has helped in all of these areas because it stresses proactive fire protection activities. The USFA and NFA have promoted the improvement of inspection/enforcement capabilities through national conferences, manuals and training courses. The USFA recently has developed a new code implementation training program and has sponsored several related reports, including Management and Enforcement of Fire Codes, Administrative Aspects of Code Enforcement, and a four-volume set prepared by the American Bar Association, Alternatives for Effective Code Enforcement and Compliance Programs at the Local Level. The four volumes are for judges, prosecutors, local officials and code officials. To promote effective code enforcement, the USFA sponsored numerous meetings for city managers, fire chiefs and state governors. The NFA offers several courses which cover inspection and plan review, including "Fire Prevention Specialists I and II," and "Plans Review for Inspectors."

Recommendation

45.. The commission recommends that, as the model code of the International Conference of Building Officials has already done, all model codes specify at least **a** single-station, early-warning detector oriented to protect sleeping areas in every dwelling unit. Further, the model codes should specify automatic fire extinguishing systems and early-warning detectors for high-rise buildings in which many people congregate.

Accomplishments

Significant advancement has been made in getting requirements for smoke detectors and automatic detection and suppression systems in codes. Currently, every major national model code requires smoke detectors in all new construction of dwelling units. In addition, some communities have amended these codes to require the retroactive installation of detectors in all dwelling units. The USFA and NFA have been extremely active in promoting the inclusion of requirements for detection and suppression systems in various codes. (The USFA provided significant input into the revision of the NFPA sprinkler standard [#13] and the development of the standard for residential sprinklers [#13D]). The USFA and CFR have conducted research into the effectiveness of such systems, held a number of conferences and workshops on these subjects, and prepared the National Directory of Automatic Suppression Systems and National Directory of Auto-

matic Detection and Remote Alarm Systems, as well as numerous other reports which support the improvement of model codes in these areas. There are still many communities which do not have adequate codes for detection and suppression systems, especially for high-risk occupancies.

2.12 Chapter 12 Transportation Fire Hazards

Recommendation

46.. The commission recommends that the National Transportation Safety Board expand its efforts in issuance of reports on transportation accidents so that the information can be used to improve transportation fire safety.

Accomplishments

The NTSB currently performs this function, for example, smoke detectors are required now in the cabins of all commercial passenger aircraft.

Recommendation

47.. The commission recommends that the Department of Transportation work with interested parties to develop a marking system, to be adopted nationwide, for the purpose of identifying transportation hazards.

Accomplishments

The DOT placard system has been developed and implemented.

Recommendation

48.. The commission recommends that the proposed National Fire Academy disseminate to every fire jurisdiction appropriate educational materials on the problems of transporting hazardous materials.

Accomplishments

The NF'A offers on-campus courses in "Chemistry of Hazardous Materials" and "Hazardous Materials Tactical Considerations," and has developed hand-off hazardous materials training packages for use by state and local fire departments. However, there are still many fire service personnel (especially volunteers) who need, but have not received, such training. The NPA hazardous materials training program is expected to be expanded over the next three years as a result of the Superfund reauthorization legislation.

Recommendation

49.. The commission recommends the extension of the CHEMTREC system to provide ready access by all fire departments and to include hazard control tactics.

Accomplishments

The CHEMTREC system has been expanded to provide increased information and assistance to fire departments. There is now a need to provide hazardous materials information using terminology that can be understood easily by fire service personnel (i.e., plain english).

Recommendation

50.. The commission recommends that the Department of the Treasury establish adequate fire regulations, suitably enforced, for the transportation, storage and transfer of hazardous materials in international commerce.

Accomplishments

Regulations for the transportation, storage and transfer of hazardous materials, which would affect international commerce, have been implemented, but by the DOT and EPA rather than the treasury department.

Recommendation

61.. The commission recommends that the Department of Transportation set mandatory standards that will provide fire safety in private automobiles.

Accomplishments

The DOT established a burn rate test requirement for interior materials in automobiles. Most of the fire safety standards for automobiles arevoluntary.

Recommendation

62.. The commission recommends that airport authorities review their fire fighting capabilities and, where necessary, formulate appropriate capital improvement budgets to meet current recommended aircraft rescue and fire fighting practices.

Accomplishments

The intent of this recommendation has been accomplished by the DOT.

Recommendation

53.. The commission recommends that the Department of Transportation undertake a detailed review of the Coast Guard's responsibilities, authority and standards relating to marine fire safety.

Accomplishments

This is an on-going Coast Guard effort.

Recommendation

54.. The commission recommends that the railroads begin a concerted effort to reduce rail-caused fires along the nation's rail system.

Accomplishments

Progress has been made in this area through the use of hot-box detectors and spark arresters.

Recommendation

55.. The commission recommends that the Urban Mass Transportation Administration require explicit fire safety plans as a condition for all grants for rapid transit systems.

Accomplishments

Grants for rapid transit systems now include fire safety plan requirements.

2.13 Chapter 13 Rural Fire Protection

Recommendation

56.. The commission recommends that rural dwellers and others living at a distance from fire departments install early-warning detectors and alarms to protect sleeping areas.

Accomplishments

The USFA and other fire organizations have conducted extensive public education programs to encourage the use of detection and warning systems in rural areas.

Recommendation

57.. The commission recommends that U.S. Department of Agriculture assis-

tance to community fire protection facilities projects be contingent upon an approved master plan for fire protection for local jurisdictions. Accomplishments

This recommendation has not been implemented by the Department of Agriculture. However, the USFA has developed and promoted a master planning process designed especially for small towns and rural areas, the Basic Guide for Fire Prevention and Control Master Planning.

2.14 Chapter 14 Forest and Grassland Fire Protection

Recommendation

58.. The commission recommends that the proposed U.S. Fire Administration join with the U.S. Forest Service in exploring means to make fire safety education for forest and grassland protection more effective.

Accomplishments

The USFA has been working with the U.S. Forest Service since 1975, and an active program is underway. Furthermore, the USFA is a member of the National Wildfire Coordinating Group.

Recommendation

59.. The commission recommends that the Council of State Governments should develop model state laws relating to fire protection in forests and grasslands.

Accomplishments

This recommendation has not been initiated.

Recommendation

60.. The commission urges interested citizens and conservation groups to examine fire laws in their respective states and to press for strict compliance.

Accomplishments

Individuals and organizations have been performing this function (lobbying for fire-safe roofing laws, for example). However, these efforts have not been coordinated at the national level.

Recommendation

61.. The commission recommends that the U.S. Forest Service develop the methodology to make possible nationwide forecasting of fuel build-up as a guide to priorities in wildland management.

Accomplishments

The USFS has an active fuel management program in progress which includes pre-deployment of resources as a function of fuel loading, weather and other factors.

Recommendation

62.. The commission supports the development of a National Fire Weather Service in the National Oceanic and Atmospheric Administration and urges its acceleration.

Accomplishments

This recommendation has been accomplished; the National Fire Weather Service is operational.

2.15 Chapter 15 Fire Safety Education

Recommendation

63.. The commission recommends that the Department of Health, Education and Welfare (now the Department of Health and Human Services) include in accreditation standards fire safety education in the schools throughout the school year. Only schools presenting an effective fire safety education program should be eligible for any federal financial assistance.

Accomplishments

Some states and local school districts have adopted requirements for fire safety education. However, it is not known if the Department of Education has established accreditation standards.

Recommendation

64.. The commission recommends that the proposed U.S. Fire Administration sponsor fire safety education courses for educators to provide a teaching cadre for fire safety education.

Accomplishments

The USFA has actively promoted fire safety education through conferences and preparation of manuals and materials. For example, the USFA prepared the Public Fire Education Planning Manual, and the report, Young Children as New Targets for Public Fire Education. The NFA conducts training courses to help in working with educators and the school system, for example, "Introduction to Fire Safety Education" and "Advanced Fire Safety Education."

Recommendation

66.. The commission recommends to the states the inclusion of fire safety education in programs for future teachers and the requirement of fire safety knowledge as a prerequisite for teaching certification.

Accomplishments

The degree to which states have implemented this recommendation is unknown, which probably means that little has been accomplished.

Recommendation

66.. The commission recommends that the proposed U.S. Fire Administration develop 'a program, with adequate funding, to assist, augment, and evaluate existing public and private fire safety education efforts.

Accomplishments

Since 1975, the USFA has conducted a program to support local public education activities. The extent of this program has varied according to the extent of funding available each fiscal year. This program has included the general promotion of public education efforts by local agencies through meetings (e.g., the "Public Fire Education Planning Conference"); preparation of support materials (e.g., Media Ideas Workbook); and provision of limited financial support through the "Public Education Assistance Program" (PEAP) which was offered during the period 1979-1981 and had to be discontinued as a result of funding reductions. The USFA is sponsoring the Community Volunteer Fire Prevention Program to promote the cooperation of local fire departments and citizen groups in conducting community

fire prevention, education and protection programs.

Recommendation

67.. The commission recommends that the proposed U.S. Fire Administration, in conjunction with the National Advertising Council and National Fire Protection Association, sponsor an all-media campaign of public service advertising designed to promote public awareness of fire safety.

Accomplishments

The USFA has conducted a number of national campaigns (e.g., the national smoke detector installation and maintenance campaign). In fiscal year 1985, a renewed public education and awareness effort was initiated as a result of additional Congressional support, An all-media campaign conducted in conjunction with the National Advertising Council has not been accomplished. However, such a program still is being pursued.

Recommendation

68.. The commission recommends that the proposed U.S. Fire Administration develop packets of educational materials appropriate to each occupational category that has special needs or opportunities in promoting fire safety.

Accomplishments

Although packets have not been developed for specific occupations, materials for pre-school children and teachers were developed in cooperation with the Children's Television Workshop (i.e., "Sesame Street"). The award winning Sesame Street Fire Safety Program is being expanded to include older children.

2.16 Chapter 16 Fire Safety for the Home

Recommendation

69.. The commission supports the Operation EDITH (Exit Drills In The Home) plan and recommends its acceptance and implementation both individually and communitywide.

Accomplishments

The NFPA has made a significant effort in accomplishing this recommendation. The EDITH program is accepted and used widely by local fire service and community groups.

Recommendation

70.. The commission recommends that annual home inspections be undertaken by every fire department in the nation. Further, federal financial assistance to fire jurisdictions should be contingent upon their implementation of effective home fire inspection programs.

Accomplishments

Home fire inspection programs of various types are conducted by most fire departments. The USFA has encouraged the initiation of such programs through research (e.g., Project RIDFIRE and the Municipal Fire Insurance Analysis): conferences (e.g., "Dynamics of Fire Prevention"); and preparation of materials (e.g., the Edmonds [Washington] Home Survey materials). Because of the constitutional issue of eminent domain, a more direct federal role is not considered appropriate.

Recommendation

71.. The commission urges Americans to protect themselves and their families by installing approved early-warning fire detectors and alarms in their homes.

Accomplishments

As a result of vigorous programs involving the USFA, other federal agencies, national fire service organizations, and state and local fire protection agencies, smoke detectors are now in an estimated 80% of dwelling units nationwide. There is now a need for programs to ensure that these detectors are tested and maintained properly.

Recommendation

72.. The commission recommends that the insurance industry develop incentives for policyholders to install approved early-warning fire detectors in their residences.

Accomplishments

Many residential fire insurance policies now offer discounts for use of smoke detectors and residential sprinkler systems.

Recommendation

73.. The commission urges Congress to consider amending the Internal Revenue Code to permit reasonable deductions from income tax for the cost of installing approved detection and alarm systems in homes.

Accomplishments

The USFA actively pursued this recommendation. However, the IRS did not feel that the income tax could be used as a fire safety incentive. Furthermore, a sponsor was not identified for the necessary legislation. However, in a related area, California adopted a constitutional amendment excluding the value of newly installed automatic detection and suppression systems in establishing property value for taxing purposes.

Recommendation

74.. The commission recommends that the proposed U.S. Fire Administration monitor the progress of research and development on early-warning detection systems in both industry and government and provide additional support for research and development where needed.

Accomplishments

The USFA and CFR have been involved extensively in smoke detector research and development. Examples of corresponding publications include New Concepts of Fire Detection, Analysis of Fire Detector Test Methods/Performance, and Detector Sensitivity and Siting Requirements for Dwellings. The CFR is studying the detector false alarm problem in hospitals.

Recommendation

75.. The commission recommends that the proposed U.S. Fire Administration support the development of the necessary technology for improved automatic extinguishing systems that would f5nd ready acceptance by Americans in all kinds of dwelling units.

Accomplishments

As a result of USFA activities, significant progress has been made in the development and installation of residential sprinkler systems. These activities have included support of system design (e.g., Development of Low-Cost Residential Sprinkler Protection: A Technical Report, Study to Establish the Existing Automatic Fire Suppression Technology

for Use in Residential Occupancies, and Sprinkler Performance in Residential Fire Tests). In addition, the USFA was a sponsor of "Operation San Francisco" which was a series of tests of retrofitted sprinkler systems (including a residential occupancy) using advanced designs and materials. The USFA also has been active in sponsoring a number of meetings to disseminate the results of these efforts and encourage the implementation of such systems. For example, the USFA currently is sponsoring a nationwide series of sprinkler workshops in cooperation with the International Association of Fire Chiefs.

Recommendation

76.. The commission recommends that the National Fire Protection Association and American National Standards Institute jointly review the Standard for Mobile Homes and seek to strengthen it, particularly in such areas as interior finish materials and fire detection.

Accomplishments

The CFR, after considerable research, recommended guidelines for fire safety to HUD, directed primarily at interior finish, for the agency's minimum property standards for mobile homes. These guidelines were adopted and implemented.

Recommendation

77.. The commission recommends that all political jurisdictions require compliance with the NFPA/ANSI standard for mobile homes, together with additional requirements for early-warning fire detectors and improved fire resistance of materials.

Accomplishments

As a result of NFPA activities, mobile home standards now require the installation of smoke detectors in new mobile homes, as well as exiting requirements and interior flame spread restrictions.

Recommendation

78.. The commission recommends that state and local jurisdictions adopt the NFPA Standard on Mobile Home Parks as a minimum mode of protection for the residents of these parks.

Accomplishments

Many jurisdictions have adopted the NFPA standard.

2.17 Chapter 17 Fire Safety for the Young, Old and Infirm

Recommendation

79.. The commission strongly endorses the provisions of the Life Safety Code which require specific construction features, exit facilities and fire detection systems in child day care centers, and recommends that they be adopted and enforced immediately by all the states as a minimum requirement for the licensing of such facilities.

Accomplishments

Many states and communities have adopted NFPA's Life Safety Code.

Recommendation

80.. The commission recommends that early-warning detectors and total automatic sprinkler protection or other suitable automatic extinguishing systems be required in all facilities for the care and housing of the elderly.

Accomplishments

All national codes and state statutes now require such protection for congregate care facilities.

Recommendation

81.. The commission recommends to federal agencies and the states that they establish mechanisms for annual review and rapid upgrading of their fire safety requirements for facilities for the aged and infirm to a level no less stringent than the current NFPA Life Safety Code.

Accomplishments

The code development process provides for the accomplishment of this recommendation.

Recommendation

82.. The commission recommends that the special needs of the physically handicapped and elderly in institutions, special housing and public buildings be incorporated into all fire safety standards and codes.

Accomplishments

This recommendation must be accomplished by standard-setting and code organizations. The USFA has analyzed these problems (e.g., Fire and Life Safety for the Handicapped) and has encouraged improvement of standards and codes and adoption by state and local governments. The NFPA has addressed the special needs of congregate care facilities in the Life Safety Code (101). If federal funds are involved, such facilities must meet the codes and standards.

Recommendation

83.. The commission recommends that the states provide for periodic inspection of facilities for the aged and infirm, either by the state fire marshal's office or by local fire departments, and also require approval of plans for new facilities and inspection by a designated authority during and after construction.

Accomplishments

Many states and local jurisdictions now have such requirements. The USFA has encouraged the adoption of these requirements.

Recommendation

84.. The commission recommends that the National Institute of Standards and Technology develop standards for the flammability of fabric materials commonly used in nursing homes, with a view to providing the highest level of fire resistance compatible with the state-of-the-art and reasonable costs.

Accomplishments

The CPSC, with technical support from CFR, has established several standards under the Flammable Fabrics Act that are also applicable to the needs of nursing homes. In addition, CFR developed the Fire Safety Evaluation System for health care facilities that aids in the objective fire safety evaluation of such facilities by both the fire service and owner/operators.

Recommendation

85.. The commission recommends that political subdivisions regulate the location of nursing homes and housing for the elderly and require that fire alarm systems be tied directly and automatically to the local fire department.

Accomplishments

This recommendation would be implemented through local zoning regulations and codes. Many communities now require central station monitoring of automatic detection and suppression systems in these occupancies.

3.18 Chapter 18 Research for Tomorrow's Fire Problem

Recommendation

86.. The commission recommends that the federal government retain and strengthen its programs of fire research for which no non-governmental alternatives exist.

Accomplishments

Fire research programs have been conducted by the USFA and CFR, but not to the extent envisioned in America Burning because of budget limitations. For example, the commission recommended an annual research budget (USFA and CFR) of $33,250,000 (1973 dollars). The largest budget ever given to the USFA for all programs was approximately $24 million.

Recommendation

87.. The commission recommends that the federal budget for research connected with fire be increased by $26 million.

Accomplishments

See recommendation 86.

Recommendation

88.. The commission recommends that associations of material and product manufacturers encourage their member companies to sponsor research directed toward improving the fire safety of the built environment.

Accomplishments

Associations and manufacturers have increased fire research activities, for example, the Chemical Manufacturers Association, The Society of the Plastics Industry, Carpet and Rug Institute, Tobacco Institute, and Concrete and Masonry Institute. These associations also have sponsored research at such private laboratories as the Southwest Research Institute, Factory Mutual and Underwriters Laboratories.

2.19 Chapter 19 Federal Involvement

Recommendation

89.. The commission recommends that the proposed U.S. Fire Administration be located in the Department of Housing and Urban Development.

Accomplishments

Originally, the USFA was in the Department of Commerce and was moved to the Federal Emergency Management Agency in 1979.

Recommendation

90.. The commission recommends that federal assistance in support of state and local fire service programs be limited to those jurisdictions complying with the National Fire Data System reporting requirements.

Accomplishments

The federal assistance funding categories included in the commission's pro-

posed budget were not included in subsequent U.S. Forest Service funding. Therefore, this recommendation is not applicable.

Section III
Summary of Accomplishments

The recommendations which have been fully or substantially accomplished generally have been associated with the agencies created as a result of America Burning or those with a public safety mission such as the Consumer Product Safety Commission.

Conversely, those recommendations not accomplished were generally the responsibility of agencies not primarily concerned with fire safety, for example, the U.S. Department of Agriculture, Department of Education and the Council of State Governments.

3.1 Recommendations Generally Accomplished

The recommendations that are fully or practically accomplished are listed below:

1 The commission recommends that Congress establish a U.S. Fire Administration to provide a national focus for the nation's fire problem and to promote a comprehensive program with adequate funding to reduce life and property loss from fire. (Note; The USFA was established, but never funded at the recommended levels.)

18.. In the administering of federal funds for training or other assistance to local fire departments, the commission recommends that eligibility be limited to those departments that have adopted an effective affirmative action program related to the employment and promotion of members of minority groups.

20.. The commission recommends the establishment of a National Fire Academy to provide specialized training in areas important to the fire services and to assist state and local jurisdictions in their training programs. (Note: The NFA was established, but never funded at the recommended levels.)

42.. The commission recommends that the proposed National Fire Academy develop short courses to educate practicing designers in the basis of fire safety design.

45.. The commission recommends that, as the model code of the International Conference of Building Officials has already done, all model codes specify at least a single-station early-warning detector oriented to protect sleeping areas in every dwelling unit. Further, the model codes should specify automatic fire extinguishing systems and early-warning detectors for high-rise buildings in which many people congregate.

46.. The commission recommends that the National Transportation Safety Board expand its efforts in issuance of reports on transportation accidents so that the information can be used to improve transportation fire safety.

47.. The commission recommends that the Department of Transportation work with interested parties to develop a marking system, to be adopted nationwide, for the purpose of identifying transportation hazards.

49.. The commission recommends the extension of the CHEMTREC system to

provide ready access by all fire departments and to include hazard control tactics.

50.. The commission recommends that the Department of the Treasury establish adequate fire regulations, suitably enforced, for the transportation, storage and transfer of hazardous materials in international commerce.

52.. The commission recommends that airport authorities review their fire fighting capabilities and, where necessary, formulate appropriate capital improvement budgets to meet current recommended aircraft rescue and fire fighting practices.

55.. The commission recommends that the Urban Mass Transportation Administration require explicit fire safety plans as a condition for all grants for rapid transit systems.

62.. The commission supports the development of a National Fire Weather Service in the National Oceanic and Atmospheric Administration and urges its acceleration.

77.. The commission recommends that all political jurisdictions requirecompliante with the NFPA/ANSI standard for mobile homes, together with additional requirements for early-warning fire detectors and improved fire resistance of materials.

80.. The commission recommends that early-warning detectors and total automatic sprinkler protection or other suitable automatic extinguishing systems be required in all facilities for the care and housing of the elderly.

81.. The commission recommends to federal agencies and the states that they establish mechanisms for annual review and rapid upgrading of their fire safety requirements for facilities for the aged and infirm to a level no less stringent than the current NFPA Life Safety Code.

3.2 Recommendations Not Accomplished

The recommendations without any significant accomplishments are listed below:

6.. The commission recommends that the National Institutes of Health administer and support a systematic program of research concerning smoke inhalation injuries.

11.. The commission recommends that federal grants for equipment and training be available only to those fire jurisdictions that operate from a federally approved master plan for fire protection.

28.. The commission urges the Joint Council of National Fire Service Organizations to sponsor a study to identify shortcomings of fire fighting equipment and the kinds of research, development or technology transfer that can overcome the deficiencies.

40.. The commission recommends to schools giving degrees in architecture and engineering that they include in their curricula at least one course in fire safety. Further, we urge the American Institute of Architects, professional engineering societies and state registration boards to implement this recommendation.

41.. The commission urges the Society of Fire Protection Engineers to draft model courses for architects and engineers in the field of fire protection engineering.

57.. The commission recommends that U.S. Department of Agriculture assistance to community fire protection facilities projects be contingent upon an approved master plan for fire protection for local jurisdictions.

59.. The commission recommends that the Council of State Governments undertake to develop model state laws relating to fire protection in forests and grasslands.

73.. The commission urges Congress to consider amending the Internal Revenue Code to permit reasonable deductions from income tax for the cost of installing approved detection and alarm systems in homes.

87.. The commission recommends that the federal budget for research connected with fire be increased by $26 million.

89.. The commission recommends that the proposed US. Fire Administration be located in the Department of Housing and Urban Development.

90.. The commission recommends that federal assistance in support of state and local fire service programs be limited to those jurisdictions complying with the National Fire Data System reporting requirements.

3.3 Recommendations Partially Accomplished

All of the remaining recommendations fall into this category. Each recommendation has been accomplished to some degree, but additional effort is required. In some cases, the recommendation may never be accomplished fully because it is on-going and general in nature.

In many cases, recommendations could not be completed because of a lack of resources, especially considering that the USFA, NFA and CFR were never funded at anywhere near the levels recommended by the commission.

3.4 Conclusions

The accomplishments discussed in this report are truly significant. In fact, 79 of the 90 recommendations have been accomplished to some degree, and the consequences of these accomplishments have had a major impact on the protection of life and property. For example:

• The annual number of fire deaths has decreased by 23% from 1975 to 1985 (an average annual reduction of 1900 deaths). On a cumulative basis, this reduction means that an estimated 6,900 lives have been saved between 1975 and 1985. Fire fighter deaths also have been reduced significantly.

• Even greater reductions in fire deaths have been achieved within special categories. Clothing fire deaths have fallen by 73% over the period 1968-1983. In addition, children's clothing fire deaths have dropped by 90%.

• The number of fires reported to fire departments has decreased by approximately 20% over the period 1975-1985. The number of reported fires has been decreasing even though the population is increasing which means that the number of fires is declining even on a per capita basis.

These achievements demonstrate what has been accomplished since America Burning was published, and the potential for what could be accomplished if the remaining recommendations were substantially completed.

* U.S. GOVERNMENT PRINTING OFFICE: 1990 - 7 2 4 - 1 5 6 / 2 0 4 3 0

www.ingramcontent.com/pod-product-compliance
Lightning Source LLC
Chambersburg PA
CBHW081129170526
45165CB00008B/2601